D0780125

Enterprising Innovation

Dedicated to: Bob and Vera

Enterprising Innovation: An Alternative Approach

Veronica Mole and Dave Elliott

Frances Pinter (Publishers), London

First published in Great Britain in 1987 by
Frances Pinter (Publishers) Limited
25 Floral Street, London WC2E 9DS

British Library Cataloguing in Publication Data
Mole, Veronica
Enterprising innovation: an alternative approach.
1. Technological innovations—Social aspects
I. Title II. Elliott, Dave
306'.46 T173.8

ISBN 0-86187-577-X

Typeset by Rapidset & Design Ltd., London WC1
Printed by Biddles of Guildford, Ltd.

Contents

List of Figures

List of Abbreviations

AUEW	Associated Union of Engineering Workers
BTG	British Technology Group
CAD	Computer-Aided Design
CAITS	Centre for Alternative Industrial and Technological Systems
CAM	Computer-Aided Manufacture
CCD	Community Construction Design
DoI	Department of Industry
ECSC	Energy Conservation and Solar Centre
ERR	Earth Resource Research
FAST	Forecasting and Assessment in the Field of Science and Technology
GLC	Greater London Council
GLEB	Greater London Enterprise Board
IT	Information Technology
ITDG	Intermediate Technology Development Group
LEEN	London Energy and Employment Network
LIN	London Innovation Network
LIS	London Industrial Strategy
LNTN	London New Technology Network
MAP	Microelectronics Applications Project
MoD	Ministry of Defence
NATTA	Network for Alternative Technology and Technology Assessment
NEDO	National Economic Development Office
NIESR	National Institute of Economic and Social Research
OECD	Organisation for Economic Co-operation and Development
PEDNEL	Product and Employment Development Agency for North and East London
R and D	Research and Development

SCEPTRE	Sheffield Centre for Product Development and Technological Resources
SERA	Socialist Environment and Resources Association
SPRU	Science Policy Research Unit
UDAP	Unit for the Development of Alternative Products
WMCC	West Midland County Council
WMEB	West Midland Enterprise Board

Acknowledgements

Thanks to those who have been helpful and encouraging during the writing of this book, particularly Melanie Corbishley and Sally Paffard for their never-ending support.

Special thanks to Dr Margaret Bruce in the Department of Management Sciences at UMIST for her constructive criticism and practical help, and to Peter Moulson (formerly of Frances Pinter Publishers Ltd) for his encouragement and assistance.

Introduction

Technological innovation is widely believed to hold the answer to many of our economic and social problems. But what sort of technology should we be focusing on and how do we decide? The 1970s saw many critiques of the social and environmental problems that, it seemed, rapid deployment of technology had produced and a consequent call for the social control of technology. In the 1980s, with commitment to innovation even more firmly established, is there any way in which we can direct technology to more socially and environmentally appropriate ends? This book explores some of the new ideas on socially-directed innovation that have grown out of the radical initiatives mounted by several local government agencies in the UK, and attempts to relate them to the wider question of what role innovation should have in society and what type of technology we should aim for.

Innovation

Technological innovation is widely perceived as a panacea for economic prosperity. World-wide there is a flurry of activity aimed at stimulating the development and deployment of new technologies of various kinds—in the expectation that this will usher in a new economic boom. Private companies allocate significant capital to ever more esoteric research and development projects, while governments, whether interventionist-minded or not, have found it necessary to allocate significant public funds to research and the support of innovation generally. In some cases, this has meant setting up special collaborative projects—like the British Advanced Information Technology Programme (Alvey 1982). In some others, it has meant trying to enhance the 'spin-off' from government-funded military projects or target certain lines of technological development by selective military (or civil) equipment procurement policies.

In general, the 'mission-oriented' approach, with large-scale centrally directed projects is seen as the most effective way of achieving well-defined targets—whether putting people on the moon or developing fifth-generation computers (Alvey 1982).

At the same time, however, a certain degree of laissez-faire is also seen as appropriate. Inventive genius—and technological creativity—perhaps cannot be centrally directed. It can flower in the most unlikely and unexpected places. Commercial considerations may not always be appropriate at the early stages of the innovation process. Academic environments are often considered as more suitable and, in particular, universities are seen as a prime source of new ideas which should be exploited; hence the current interest in Science Parks.[1]

Essentially what is being suggested is that there is a need to transfer and absorb ideas from a diffuse network of inventive sources for commercial purposes. So, along with innovation, 'technology transfer' is seen as a vital necessity. This concept is wider than that of just transferring ideas from academic to commercial use. It also covers the transfer of technology and knowledge from military-orientated research to civil use, from basic science to applied technology and indeed from 'advanced' to Third World countries.

The implications are that the first stage in the innovation cycle—invention—is relatively unproblematic in that there is likely to be a reasonable and growing pool of ideas which could be exploited. Rather, it is implied that the problem is to transfer these ideas from their source to where they can be developed and deployed successfully in commercial terms.

Closer inspection of the technological innovation process, however, indicates that the problem is far more complex. 'Invention' rarely just happens spontaneously. The next stage, moving to a prototype, is equally complex, as is the subsequent step of moving to volume production. At each stage, there are likely to be hurdles at which most potential innovations will falter. Quite apart from purely technological problems, there are the financial and institutional hurdles. For example, few investors are keen to risk money on long shots, so inventors usually find initial funding a major problem. The time-scales for the stages from idea to prototype, much less to manufacturable product, are too long for many investors.

In recent years, some funding agencies have tried to fill this gap, although they have tended to be cautious, choosing projects which appeared likely to be commercially successful. Clearly this approach has its limitations. Novel ideas may get ignored. More generally, commercial viability is not always the same as social viability; the competitive market mechanism is not always the ideal arbiter of social value. As Harvey Brooks wrote in an OECD report: 'competition sometimes rewards the least responsible corporate behaviour, while exacting a large social cost in regulating and alleviating the disbenefits' (Brooks *et al.* 1971).

Socially-directed Innovation

Concerns like this led, in the 1970s, to attempts to regulate the development of technology so as to avoid social and environmental problems, via technology assessment, environmental impact studies, and so on. However, there are a number of problems with these attempts at subjecting technology to social control. It is hard to act early enough to have an impact, since there is usually insufficient information, but it is difficult to leave assessment until later because new technologies may then become established with an economic momentum of their own. At worst, it is possible to wait until a technology is deployed or is about to be deployed, before it can be effectively assessed. But, this can lead, in the case of negatively assessed options, to possible social and even political confrontations as more and more people, previously unaware of the issue, begin to take sides.

With the worsening economic climate of the 1980s, technology assessment and environmental impact analyses, quite apart from public participation in technological decision-making, have been seen more and more as a luxury that cannot be afforded. If assessments are to be made, it must be to decide which option is more likely to be most commercially successful, with only passing reference to environmental and social considerations. The emphasis has moved from *regulation* to *stimulation*—finding ways to identify and promote successful new products.

Interestingly, however, those still interested in the social control of technology now argue that one way to try to avoid undesirable technologies emerging is to seek to influence the *early stages* of the innovation process. So they too have seen the stimulation process as strategically important—albeit for slightly different purposes. For them, stimulation of desired technologies (and presumably repression of undesired options) is a species of political and economic intervention. Rather than trying to assess proposed new projects from the 'outside', social control can, it is agreed, be introduced effectively as *part* of the funding support process by *public* agencies operating at the early stage of the innovation process.

Given that commercially orientated agencies seem loath to risk investing in new ideas, the field is relatively open for public agencies—such as national or local government bodies—who can therefore, at least in principle, apply non-commercial criteria in selecting which areas to support. Clearly this is essentially a political prescription—based on the belief that elected governments can and should apply broader criteria to public investments than those emerging from the market-place, for example via the research they support and the procurement policies they pursue. Given that some governments seem to equate the 'national interest' with the interests of the major corporations (whether publicly or privately owned) there is obviously likely to be some conflict over political control if this prescription is acted on fully.

There is nothing new in the idea of using state funding as means of promoting politically chosen technological policies: that is what many governments do. What is new is that, as we shall see, radicals of various hues have now attempted to intervene in this process more effectively—introducing a wide range of social, environmental and political criteria into the decision-making process.

In terms of innovation process what is being argued fundamentally is that innovation should be need-led rather than just profit-led, and that not *all* innovation is necessarily desirable in social terms and that it is possible and necessary to be both selective *and* democratic in one's approach to innovation, with innovation being seen as part of social as well as purely technological policy.

The British Experience

In the past five years several experiments along these lines have been mounted by radical local government agencies in the UK. Public funds, raised from taxes on property (the 'rates') have been deployed to stimulate politically chosen lines of technological development, with employment creation being one of the major aims.

There has been a concerted attempt to link the technological resources of local colleges and polytechnics to meet the strategic social, economic and employment needs of the community—as a clear alternative to the practices adopted in Science Parks, where academic expertise is channelled to meet commercial requirements (for an overview of local authority technology policies in the UK, see Marvin 1986).

In part, these local projects and programmes have come about because central government in the UK has adopted a non-interventionist economic policy. Clearly, then, the radical approach has drawn on socialist thinking concerning the need for planned economic intervention. But it is not simply a conventional state-corporatist approach—technocratic planning from above. Attempts have been made to involve community networks, trade union organisations, user groups and so on in the planning process and thereby to identify and react to social needs.

These initiatives raise a whole spectrum of practical and theoretical issues and problems. Can 'needs' be articulated effectively? Can they be used to guide the innovation process? Can 'socially needed' products also be commercially viable given the continued existence of a competitive market mechanism? Can this approach transcend the limits of conventional (and currently beleaguered) 'welfare socialism'? And finally, what are the longer-term options on innovation?

This book attempts to explore these issues, drawing on the UK experience. Our primary concern is to map out the theoretical justifications for—and problems of—a process of socially-directed innovation.

A word about the structure of this book. There are three distinct phases. The first part consists of a review of some of the relevant theories of innovation coupled with a critique of current practices and approaches, written by Veronica Mole. We then move on to a case study of the radical approach adopted by the Greater London Council (GLC). Finally we move to a more general analytical approach.

Although we look critically at innovation, we do not attempt to provide an assessment of the social impacts of new technologies or a review of technological options except in very broad terms. In part this is because such themes have been well covered elsewhere, but in addition we felt that our emphasis should be more on the social process of innovation. The word 'impact' implies a rather one-way process; instead we see it as an interactive process.

Certainly technology can have disastrous accidental or unexpected interactions with society and the environment. But our emphasis is more on how the consciously chosen outcomes are selected, rather than looking at their impacts. For some people the effects—chosen or otherwise—of the interaction between technology and society are so worrying that they would prefer some sort of moratorium, or at least a slowdown, in the rate of innovation and change in its direction.

This position need not be seen as Luddite (and indeed the Luddites were only trying to shape the development of technology in their interests). All innovations are not necessarily beneficial and our capacity to select appropriate lines of development may be no match for the rapidity with which new technologies are emerging.

But the point is that while some specific developments may be inappropriate, one cannot deal with situations piecemeal: a new comprehensive social process for assessment and direction is needed that would amount to a whole new approach to innovation and the role of technology in society.

The first fragile shoots of such an approach can be discerned in some of the examples described in this book. While it would be unwise to suggest that many of the complex problems of 'social control' have been solved, we do feel that the experiments we review are worth careful consideration by those interested in how technology and society should develop.

This book is based on some ideas developed by Dave Elliott for an Open University course (T362, *Design and Innovation*, Open University, 1986) which contains a case study on the Greater London Enterprise Board (GLEB). Dave Elliott spent one year as a part-time consultant working in the Technology Division at GLEB on the establishment of the London Energy and Employment Network (LEEN). This and his involvement with LEEN gave an inside view, although the book is based on published material. The investigations into GLEB and GLC were continued by Veronica Mole, from 1983, as part of her doctorate work and as a researcher attached to the Open University Technology Policy Group. Parallel work going on in the Open University Technology Policy Group on Science Parks and technological innovation has also fed into this study.

However, the final responsibility for what appears here rests with the authors. We are grateful to the Open University Faculty of Technology for providing the Technology Policy Group with funds to allow Veronica to work on this book and to our colleagues at the Open University who have provided advice and support, particularly Paul Quintas.

Notes

1. Science Park is the name given to industrial development areas that are situated adjacent to University campuses with the aim of industrial/academic liaison on leading edge technologies (see Lowe 1985).

PART 1: The Innovation Process

Introduction

Technology and Society

Technology permeates all modes of existence in modern Western societies, illustrated by the fact that almost all human activity is reliant upon some form of technology—the early morning cup of tea, domestic routines, work patterns and organisation, entertainment and so on.

The history of modern industrialised societies is related to developments in technology. The importance of steam power to the Industrial Revolution and the development of machines such as the spinning-jenny and Crompton's mule are basic to the passage from an agrarian to industrial society. Likewise, the post-war growth in new technological areas of electronics, synthetic materials, oil and consumer durables has laid the foundations for the patterns of modern living. Technology has brought great changes in work patterns, communications, the consumption of products and services. In sum, the structure of our society is closely related to the state of technological knowledge and development.

Technological determinism, that is, the 'machines make history' view, is based on the belief that technological change causes social change. This cause-and-effect relationship between technology and society implies that technology is autonomous and socially neutral; the path of technological development is unilinear and related to objective factors such as technical requirements for economic growth. Within this perspective, technology is conceived as 'machinery' as the physical artefacts. The products, processes and the physical and organisational requirements for efficient production comprise the units of analysis or methodological base from which to understand the innovation process. Thus, the context for much of the research on technological innovation is orientated towards the identification of the technical and organisational factors relating to commercially successful product or process innovation, for example, research and development (R and D) activities and so on. In effect, this serves to promote a prescriptive approach to innovation management. The belief that innovation leads to economic growth acts as a powerful justificatory argument for more innovation and increasingly sophisticated technologies to ensure competitiveness in world markets. The adoption of competitive commercial criteria results in the selection of specific routes for technological development.

This strictly economic view of technology is matched by a more general ideological view, in which technology is equated with human progress. The level of development of machines, materials and tools employed in the creation and maintenance of the human environment serves as an indicator of technological advance and, by implication, of social progress, for example, the association of information technology (IT) with the information- or knowledge-based society (Bell 1974). The allocation of public funds, such as, via the British Programme for Advanced Information Technology (Alvey 1982), to the development of new technologies, reflects this dual economic/ideological function of technology. In the UK this has taken the form, under the present Conservative government, of support for selected research programmes aimed at the commercial development of 'high-technology' industries: microelectronics, IT, biotechnology and defence projects.

However, the focus on the economic/ideological function of technology and its realisation within products and processes provides a limited explanation of the nature and role of technology in society. In this view, the potential social impacts and unforeseen side-effects of technological innovations are a separate issue, the inevitable costs of the technological project (the 'technical fix' refers to the technical solutions that emerge to cure these side-effects: Weinberg 1966). Yet the social implications of technological change comprise the larger context of technology in society.

In sum, the criticism of modern technology has centred around the exploitative use of natural resources, high pollution rates, high capital costs and the danger to public health and safety. One argument is, for example, that the risks involved from radioactive leaks from nuclear power stations are too high to be offset by the social benefits of allegedly cheaper forms of electricity. In addition, the implications of technological innovations for work organisation—worker alienation, deskilling and technological unemployment—have received critical attention. A central point of these criticisms of modern technologies is that there is a mismatch between technology and society. The increasing sophistication of our technological environment at the expense of employment and the natural environment, and the development of ever-deadlier weapons of war and mass destruction have lead to doubts as to whether the identification of technological progress with human progress is justified. What technology can provide and what technology actually does provide can result in massive contradictions. Mike Cooley, a leading member of the Lucas Shop Stewards' Combine Committee and now the Director of the Technology Division of the Greater London Enterprise Board (GLEB) put this most succinctly, when he wrote: 'we have a level of technological sophistication such that we can design and produce Concorde, yet in the same society we cannot provide enough simple heating systems to protect old-age pensioners from hypothermia' (Cooley 1985, p. 19). This is one example among many others.

The Social Construction of Technology

In this book, we focus upon the innovation process as the core to the problem of technology. Central to our argument is the understanding of technological innovation as a social and political process. From this perspective, technological development is neither autonomous nor unilinear; it is the product of a social process of selection which reflects economic and political priorities of those with the power to choose.

The 'social construction of technology' (see Mulkay 1979; Pinch and Bijker 1984) is an embryonic research area which attempts to relate the insights of the sociology of scientific knowledge to technology. The sociology of scientific knowledge has sought to reveal the construction of scientific knowledge by illustrating its status as one among different and competing knowledge cultures. These concepts can also be related to 'technology as social construction'. An attempt is made to demonstrate that technological artefacts reflect the social process of their design and development. Pinch and Bijker (1984, p. 421), in a study of the social construction of the bicycle argue that 'technological artefacts are culturally constructed and interpreted—not only that there is flexibility in how people think of or interpret artefacts, but also that there is flexibility in how artefacts are designed'. For example, a car can be variously interpreted as a status symbol, a convenient mode of transport or a pollutant of the environment. The design configurations of the vehicle can reflect these different interpretations in the form of a sportscar, a family saloon and an electric-powered vehicle, like the Sinclair C5, respectively.

The social conception of technology emphasises that the design, innovation and implementation of technologies exhibit a number of potentially different directions of development. Thus, there are always alternative designs of products and processes and alternative solutions to technological problems. For example, the problem of energy supply can be met in a number of different ways: nuclear technology, solar technology and wind-power technology. The design or form of solar panels can differ according to particular user requirements and environmental conditions. The important question that emerges from this is why certain technological designs are preferred to others.

The social groups involved in the stages of the design and innovation process are identified as crucial to the end-product; technologies embody these social relations. A proponent of this view, Johnston (1985, p. 381) writes:

> the shaping of technology is a social process, from selection of research projects and market targets, to the form of its introduction, degree of adaptation required by the economy and society into which it is introduced—and ultimately to the values and goals of an industrial society. These social processes and . . . these interactions are responsible for the ways in which technology augments and limits human capability.

In an article concerning the social viability of technology, Wynne (1983, p. 16) makes the point that 'the present social organisation of innovation and design is dominated by technical experts oriented toward specific priorities'. The technological environment is geared towards economic growth so that questions, such as whose interests this technology will serve, or who will be affected by it, and so on, are not asked. The needs of women and other groups often remain unrecognised and neglected by the (typically male) 'experts' (Bruce *et al.* 1984), the designers and innovators, and the 'decision-makers' who make the crucial choices between alternative solutions (Bruce 1985).

Technological determinism and the social construction of technology are competing views of the relationship between technology and society. The innovation process plays a key role in both arguments. On the one hand, it is central to continued economic growth. On the other hand, it is important to the understanding of the social construction of technology and the possibility of intervention in the design and development process that will enable the opportunity to create alternative solutions to technological problems.

A point to recognise is that those concerned with alternative directions of technological development are not inherently anti-technological. Rather, alternative conceptions of technological development are based upon the inclusion of a different range of social groups, for example, the elderly, women, ethnic minorities, and so on, in the social processes which shape technology. For the 'radical' local authorities, the creation of the facilities to encourage socially-directed innovation was not a marketing exercise but an ongoing political/technological project.

We can summarise the basic argument we are developing as follows:

1. The link between technological innovation and economic growth provides a justification for the selection of routes to technological development in competitive market economies.
2. Economic models of the innovation process which concentrate upon the explanation of 'successful' (in commercial terms) innovation management tend to obscure the multi-directional nature of innovative activity.
3. The social construction of technology seeks to reveal that the 'successful' stages of technological development are not the only possible ones. At every stage, there are alternative solutions which reflect social, political and economic values. The innovation process is a social and political process that is dominated by 'experts' and it reflects the dominant interests in society.

In this book, we are concerned to investigate the nature of the innovation process and to identify the factors that affect the practical attempts (as discussed in Part Two) to direct innovative activity toward social objectives.

In Part One an attempt is made to explore the nature of the innovation process in an economic and social context. This seems a fruitful way of tackling the field of innovation studies since the literature can often appear rather abstract and removed from the very real problems associated with technological change.

In Chapter 1, the economic approach to the study of innovation is explored. The concepts of 'technology-push' and 'market-pull' which inform much of the literature on innovation are used to illustrate the theoretical basis of empirical studies of innovation. In addition, the critique of these concepts provides the theoretical justification for an alternative approach to innovation. The concept of 'technological paradigm' (Dosi 1982) is used to illustrate the nature of the technological environment in which the innovation process is located.

In Chapter 2, the theoretical underpinnings of the social construction of technology are discussed and related to the concept of 'socially useful' products. The difficulties of using the concept of 'needs' as the foundation for an alternative approach to innovation is explored. The models of innovation identified in Chapter 1 are used as a basis for the possibility of a socially-directed innovation process with references to policies adopted by 'radical' local authorities in Britain.

1 The Economic Approach to Innovation

This chapter looks at the nature of the innovation process and at the role of public agencies in the finance and support for innovation. The argument may be summarised in the following way—the identification of technological innovation with increased productivity and economic growth leads to economic analyses of the innovation process which seek to reveal the factors necessary for the development of 'successful' innovations. Within these analyses, discussion of the technology itself is marginal.

The orientation towards increased productivity, market competitiveness and the satisfaction of needs and demands through the market-place reflects the dominant values within competitive market economies. This can be illustrated by the concept of 'technological paradigm' (Dosi 1982).

The role of national and local government agencies is to support and facilitate technological innovation to ensure economic growth. The belief is that economic growth will secure the satisfaction of social needs, indirectly, by increases in the standard of living of the populace. Opponents of this view argue that the market mechanism is inadequate to ensure the satisfaction of social needs and that it is the role of public authorities to develop alternative socially-directed models of product development and innovation.

Definitions

Technological innovation refers to the process of creation, evolution and development of technological artefacts. The innovation process typically involves a series of stages ranging from the idea of invention through product design, development, production and adoption or use. Certain theorists use the term 'technological innovation' to define this whole process, Roy (1986) refers to the definition adopted by the Organisation for Economic Co-operation and Development (OECD) in the 'Frascati manual' (a document issued as a proposed standard for measuring scientific and technical activities). Here, technological innovation is described as 'the transformation of an idea into a new or improved saleable product of operational process in industry or commerce' (Roy 1986, p. 2). Within this view, innovation is a process by which new products and techniques are conceived, developed and launched.

This rather broad definition is modified within economic approaches to the study of innovation where invention is distinguished from innovation, so that Mansfield (1968, p. 83) writes that 'an invention, when applied for the first time, is called an innovation'. From this position, innovation refers to the first introduction onto the market or into social use of a new product or process. The economist, Joseph Schumpeter, regarded 'invention, the discovery of a new tool or technique, as the initial event; and innovation, the implementation of the new tool or technique, as the final event' (cited in Kelly and Kranzberg 1978, p. 2). This distinction between invention and innovation is useful as a guide to understanding the difference between the discovery of new products and processes and their eventual application; only a proportion of all inventions ever reach the point of commercial or social use. The distinction between innovation and invention is important for our purposes because it allows the possibility of posing the question why certain technologies are developed and exploited.

Within economic approaches to the study of innovation, the factors relating to the commercial success of the innovating firm—R and D capacity, management strategies, marketing techniques and macroeconomic factors within the economy as a whole—comprise the units of analysis. Thus, the emphasis is on the process of innovation rather than on the physical characteristics of the technology in question. The process of converting an invention into an innovation involves a number of phases. A separation between the terms 'science', 'technology' and 'technique' is often used to identify different points of the innovation process. 'Science' refers to the discovery of pure scientific knowledge that is not immediately applicable, while 'technology' refers to their application of knowledge to a particular problem. 'Technique' is the actual production and diffusion of this knowledge in the form of commodities. Thus, 'technology' is defined as knowledge and 'technique' as processes. For example, medical technology refers to the overall development and improvement of medical techniques—pacemakers, kidney machines, and so on.

This separation of science from technology and technique is useful for understanding the different processes involved in innovations. In many cases, a distinction is made between 'radical' and 'incremental' innovations. 'Radical' innovations refer to products and processes that result from advances in knowledge, for example, the motor car or the electric light-bulb. 'Incremental' innovations refer to the continual process of improvement of techniques, for example, since the invention of the motor car there have been substantial changes and improvements which qualitatively differentiate the modern Ford Escort from the Model T Ford. The distinction between 'radical' and 'incremental' innovations is not clear-cut and many innovations are difficult to categorise. The colour television, for example, resulted from a series of 'incremental' changes but has had radical effects. Likewise the zip fastener resulted from R and D advances in knowledge but could be seen to represent an 'incremental' innovation in terms of the development of types of fastener.

Technology-Push

The precise nature and cause of technological innovation, as a process, is subject to controversy and debate. In this section, the determinants and rate of technological innovation are discussed in terms of two opposed models. These are 'technology-push' and 'market-pull' (Langrish *et al.* 1972). This debate forms the base from which the innovation process within competitive market economies is characterised. Proponents of the 'technology-push' perspective argue that it is advances in science and technology which lead to changes in the composition of products and processes. Proponents of the 'market-pull' view hold that market demand is the major influence on innovative activity (Freeman 1979). Most of those involved in the debate tend to fall somewhere between these two extremes.

Technology-push models emphasise the central role of science in producing invention or the dependence on previous advances in technology to stimulate inventive activity. The relationship between science and technology is an interactive one. The view that technology is the embodiment of previous advances in science is an oversimplification of this relationship and is not supported by empirical evidence. Historically, there are many instances where technological knowledge has preceded scientific knowledge and has provided the base for scientific research, for example, Torricelli's demonstration of the weight of air in the atmosphere, which represented an important scientific breakthrough, grew out of his attempt to design an improved pump. Pasteur's development of the science of bacteriology grew out of his attempt to deal with the problems of fermentation and putrefraction in the French wine industry (Rosenberg 1982, p. 142). In some cases the natural trajectories, the paths of technological development (Nelson and Winter 1977) in a particular sector, identify the need for further scientific research. Rosenberg (1982, p. 148) gives the example of the aircraft industry where the turbojet first led to the creation of a new supersonic aerodynamics.

Despite the argument for an interactive relationship between science and technology, emphasis upon the supply-side and technology-push nature of innovative activity still presents a picture of a unilinear path of technological development. The view that science plus technology equals production implies a technological determinism where paths of development appear autonomous and technology develops under its own momentum. This scenario of a linear sequence is clearly questionable. In some cases a time-lag of ten to twenty years can exist between pure research and innovation; for example, the Bacon fuel cell was first observed in 1842 by Sir William Groves but it was not until the 1960s, when huge investments were put into the NASA space programme, that it was developed and produced in order to play a key role in the technology used to put the first astronaut, on the moon.

In its purest form technology-push assumes that there is not a dependence between innovative activity and economic factors (Haeffner 1973). But clearly economic factors such as market demand must influence the development of technologies and can be used as an indicator as to why certain technologies are developed and not others. Within advanced industrial economies the technological sphere is influenced by powerful economic needs and incentives. The allocation of scientific resources depends to a great extent on the perceived financial rewards to be expected by the advances in the technology. The increasing institutionalisation of research in private industrial and government-financed laboratories supports the view that the pursuit of research is largely directed and limited by economic costs and benefits.

In a study of a number of innovations in the areas of process plant, synthetic materials and electronics (Freeman 1982), it was found that the distinctive feature of these industries was that they were research-intensive. In addition these areas represented the three main sectors of fast-growing new products, identified by research undertaken during the 1960s and 1970s by the National Institute of Economic and Social Research (NIESR) and the Science Policy Research Unit (SPRU). The conclusions drawn were that there was a significant correlation between Research and Development efforts and innovative output in these sectors. This appears to support a technology-push argument. However, this is not the whole story, if one takes the example of the electronics industry then the importance of economic and institutional factors is apparent for the direction of resources to this area.

The origins of the radio industry at the beginning of this century were followed around the period of World War II by three of the most important innovations in the electronics industry: television, radar and the computer. All of these innovations were largely influenced by advances in fundamental science—solid-state physics. But factors apart from technological capability were important for their development. In the field of radar and computers the close relationship between the scientific community and industrial and military users was a key factor in the innovation process. Wartime military demand spurred on many new technical developments which, after the war, were not met by commercial demand for a fairly long period. For example, the growth of the European semiconductor industry was slack in the face of an absence of strong market demand. Throughout the period of technology-push until commercial innovations got under way, the combination of government-sponsored university and industrial Research and Development was crucial (Freeman *et al.* 1982).

Clearly, certain industrial sectors are more closely related to scientific and technological research than others. Technology-push approaches oversimplify the nature of the innovation process and tend to underrate the role of economic and institutional factors for innovative activity.

Technology-push explanations, however, are useful for understanding radical innovations. Langrish *et al.* (1972) suggest that the radical innovations in the electronics industry may be explained more fully by technology-push perspectives rather than the market-pull view. In support of this, Rosenberg (1982) argues that radical innovations—the major technological breakthroughs—signal only the beginning of a series of technological developments. Radical innovations may be seen as providing a new framework or technological trajectory which can shape subsequent research and development for many years. Concentration on technology-push or supply-side factors cannot explain incremental innovations. Once a major breakthrough delineates the direction of further research and technological improvement then demand factors are central to the continuation of innovative activity.

Empirical research findings suggest indirect relationships between science and innovation; most innovations are not a result of the direct application of basic scientific knowledge (Myers and Marquis 1969; Langrish *et al.* 1972). Rather science acts in conjunction with market demand and for ongoing innovative activities it is important for the resolution of technical problems (Gibbons and Johnston 1970; Rosenberg 1974). The conclusion that can be drawn is that while science is one of many factors that make up the innovative environment, it is not the main determinant of innovation.

When science does facilitate innovation it is in an indirect way, Langrish *et al.* (1972, p. 40) point to the ways that this can occur:

> First, curiousity oriented science, practiced largely in academic institutions, provides techniques of investigation. Second, it also provides people trained in using these techniques as well as in scientific ways of thought in general . . . Third, science enters innovation already embodied in technological form. It may be relatively rare for a piece of curiousity oriented research to generate a piece of new technology, but once this process has occurred, the technology can be used over and over again and developed into more advanced technology.

In sum, the main insight into the nature of innovative activity provided by the discussion of technology-push explanations is that the unilinear conception of science–technology–production is misleading and that science and technology as bodies of knowledge are interactive. A crucial point made by Langrish *et al.* (1972) is that when science does facilitate innovation it is in the form of personnel trained in the techniques of investigation who then establish a direction for further technological development; what Rosenberg (1982) terms a major or radical innovation which sets up a technological trajectory. The growth of institutionalised R and D confirms the perceived significance of scientific and technological research for innovative activities. The prime motive is the expected economic return on investment in the form of technological applications. Economic and institutional factors, especially government and military support, are important for the transformation of an invention into an innovation and in the support for selected paths of scientific and technological enquiry.

Market-Pull

Pure demand or market-pull models of innovation point to the recognition of a need as the main influence on innovation. Needs may be in the form of market demands, government or military requirements or social needs (Utterback 1974). Producers attempt to link technological efforts to the fulfilment of these needs.

The market is a mechanism whereby suppliers and consumers are brought together to facilitate the exchange of goods and services. Within competitive market economies, the market presupposes the existence of purchasing resources. Market 'needs' are in effect economically-backed 'demands'. The economic significance of 'demands' is the key factor in the connection between technological innovation and the market. In simple terms, the market-pull model of innovation is as follows. At a given time, the market consists of a range of goods which satisfy consumer needs and demands—the purchasing patterns of consumers reveal their desires and preferences. Movements in demand and price, as a result of, for example, growing incomes, act as indicators to producers that certain goods are more in demand, that is, there is a greater need for them. At this point the innovative process begins and successful producers will be those who can fill this demand for new and improved products. The demand-pull approach emphasises the primacy of market forces in both the capital-goods market and the consumer-goods market. Further demand may be in the form of government markets or particular forms of needs. For example, British wartime needs spurred the successful development of radar (Freeman 1982).

In the Myers and Marquis (1969) study, the analysis of 567 different innovations is an example of this 'one-directional' approach to the understanding of the innovation process, so that 'needs' and 'demands' determine suppliers' innovative activity. The main point in this argument is that it is possible to know the direction in which the market is 'pulling' inventive activity before an invention takes place. During the late 1960s and early 1970s, a number of empirical studies of industrial innovation, both in Britain and the United States, appeared to support this view. Included in these studies are the Queen's Award[1] study (Langrish *et al.* 1972), Project SAPPHO (SPRU 1971) and Project Hindsight (Isensen 1967). The main conclusions drawn from these studies is that factors relating to the market were important, as Bruce (1984, p. 41) states: 'the analysis of future markets and understanding the needs of future users and political goals were more often associated with successful innovations than were discoveries or "bright ideas" '. The close connection between the success of an innovation and the understanding of market and user needs is used as a supporting factor of the demand pull argument.

However, there are a number of difficulties in using this approach to explain the innovation process. A critique of market or demand-pull models (Mowery and Rosenberg 1979) points to their inadequacy in providing firm evidence for the conclusion that market demand is the major determinant of innovative activity. Firstly, they argue that there is an extremely wide definition of 'demand factors' used in many of the studies. Secondly, different definitions of 'needs' and 'demands' serve to make their findings ambiguous so that the grouping together of such studies as supporting the demand-pull perspective is questionable. Coupled with the theoretical and methodological weaknesses of these studies is a further point of concern; their inability actually to provide an adequate explanation of the innovation process.

As an explanation, innovation theory must include various types of innovations, both 'incremental'—the technical improvements to existing products and processes—and 'radical'—the major technological breakthroughs. A second point made by Mowery and Rosenberg relates to the 'radical' category of innovations. They contend that potential needs and demands in this area are possibly limitless but that it is problematic to see how these potential needs and demands can provide an explanation as to why at a given point in time innovation occurs. The why and when of particular technological developments and not others, together with the timing of such developments, remains unanswered. Moreover, the major technological breakthroughs which do not have any direct relationship with the conditions of the market also remain unaccounted for. Dosi (1982, p. 149) makes this point when defining 'needs':

> at one extreme, one may define them in very general 'anthropological' terms (the needs to eat, have shelter, communicate, etc.) but then they express a total indifference to the way they are satisfied and do not have any economic relevance; or, at the other extreme, 'needs' are expressed in relation to the specific means of their satisfaction, but then each cannot emerge before the basic invention to which it is related.

Put another way, in the first definition the need for transport can be satisfied by a number of different means or design configurations, such as a horse, a bicycle, a car. In the second definition of need, as in the need for a car, this could not be satisfied until a car had been invented. The same argument can be related to a whole host of 'needs', for example, for a washing machine or a computer. But one cannot express a need for a product that does not already exist.

A third point of criticism of market-pull approaches is their failure to confront the specific events within the innovation process between the recognition of demand by producers and the end result: the appearance of a new product. It would seem that technological possibilities are already in existence but that they have not previously been exploited. The question that arises is why this is the case. If it can be argued that technological innovations are dependent on 'needs' and 'demands', then the picture that emerges is one of a technology that is extremely versatile and which can be guided in any direction at any time. This results in the crude conception of technology as an essentially passive, reactive mechanism; a 'freely available black box' (Dosi 1982, p. 147).

Whilst the empirical studies reviewed by Mowery and Rosenberg found that the market is important in determining innovations, they do not produce enough evidence for the view that 'needs' and 'demands' expressed through the market are the major determinants of innovative activity.

Interactionist Models

In order to achieve a comprehensive picture of the innovation process which can explain both 'radical' and 'incremental' innovation, reliance upon technology-push or market-pull models exclusively is inadequate. The use of both models together is more helpful. In this way, innovative activity is explained by recourse to the role of technological and market factors in 'successful' innovations.

The interaction between science, technology and the market as a constant activity provides a more reasonable view of innovative activity. A basic uncertainty is present in all attempts to introduce new innovations and reliance on only one of these factors alone can lead to 'failures'. The most important conclusion from the SAPPHO study of the determinants of success and failure in industrial innovations (SPRU 1971) was that successful innovations (innovations were judged in terms of commercial success) are those that matched the technology with the market, where consumer and user requirements were understood and adequate resources were made available for research, development and the launch of the innovations. Mowery and Rosenberg (1979, p. 105) argue that: 'Both the underlying, evolving knowledge base of science and technology as well as the structure of market demand play central roles in innovation in an interactive fashion.'

Innovation is what Freeman (1982, p. 168) terms a 'coupling' process, that is, 'an idea "gels" or "clicks" somewhere at the ever-changing interfaces between science, technology and the market'. Innovation may be characterised as a response to a 'need' or 'demand'. On the other hand, it may involve existing or new scientific and technological knowledge resulting from research activity. Experimental development and design, trial production and marketing involve a process of 'matching' the technical possibilities and the market' (Freeman 1982, p. 166).

The discussions of technology-push and market-pull explanations of innovation have indicated that product and process innovations arise out of the nexus of activities between technical possibility and market demand. Technical change is seen as a cumulative process of minor improvements, incremental innovations, with only occasional major or radical innovations. Radical innovations serve to lay the base, or institute a technological trajectory, for a host of complementary technologies. Technological trajectories of development serve to indicate that innovations arise from an 'evolutionary' progression (Ray 1985) whereby technology builds on existing technology in an incremental or step-by-step manner.

The interaction between technical possibility and user needs is a dynamic process; user needs and the particular technical solutions to these needs are likely to change. Different technologies provide the solution to user problems, for example, the problem of cooking can be met in a number of different ways and with quite different technologies depending on the context of use, particular user requirements and available resources, for example, the primus stove, gas cooker and microwave oven. Within one particular trajectory of technical solution new innovations can be stimulated by the user's perception of the inadequacies of the existing technical solution on offer. The technical advances arising from user needs involve more appropriate ways of doing things.

At the level of the firm, innovations embody the technological knowledge required to produce the physical artefact, thus for a firm operating in a particular sector and market, products and processes comprise the cumulation of knowledge and perceived problem-solution factors necessary for their continued product development. Producers tend to work within existing technological trajectories; this is a point made by Rosenberg (1982, p. 129): 'Specialist producers tend to be very good at improving, refining and modifying their product. They tend to be weak in devising the new innovation that may constitute the eventual successor to their product.'

The criteria for the assessment of product and process innovations developed by Ray (1985) indicates that for innovations to be successful they need to complement surrounding technologies, correspond with user needs, have some advantage over competing technologies and be offered at an appropriate price. The evaluation of products and processes in these terms offers an explanatory device for successes and failures. For example, the Sinclair C5 electric-powered car offered a low-cost form of transport but did not correspond with other user needs of safety or speed. It was not in harmony with existing technologies and did not have advantage over competing technologies. Concorde corresponds with user needs for faster air travel and in this way has some advantages over competing technologies, but the price precludes its use for mass air travel. The implications of Ray's criteria for innovations lie in the tendency for technologies to develop along existing trajectories in an incremental fashion.

Product design, redesign and readaptation comprise a large part of innovation activity once a trajectory has been established. Reinnovation and adaptation has been a key factor within the car industry which has a history of improved design configurations, engineering specifications, and safety improvements within a model range.

Models of the Innovation Process

The linear-sequential model of the innovation process shown in Figure 1 has been criticised for its tendency to posit a one-directional sequence within a set time-scale. Innovation within this model is essentially static, that is, once the innovation has occurred the model does not allow room for reinnovation or readaptation. The linear sequence suggests that innovation precedes diffusion but Rothwell and Gardiner (cited in Walker 1986) found from their study of the hovercraft industry that it can be the case that diffusion precedes innovation in the sense of reinnovation and readaptation.

A model of the innovation process needs to recognise the interrelationships between different phases within the process together with the ability to present a dynamic process that is constantly changing through interaction with the wider technological environment and the market.

Walker (1986) characterises a model of innovation in terms of the 'innovation spiral' shown in Figure 2. This attempts to link the three phases together within a sequential series of events that exhibits a cyclical progression. Descriptively, the cycle identifies an invention and development stage which is affected by technology and engineering, a diffusion stage which is affected by market and demand considerations, and the stage of maturity of an innovation which leads to either displacement, adaptation in terms of redesign and product stretch or reinvention which brings the innovation full circle.

Within this model a key role is given to the importance of design and redesign of products and processes within innovation. Walker (1986, p. 34) argues that design lies between technology-push and market-pull: 'Design then occupies the centre of the innovative process because it converts general ideas and needs into specific objects.'

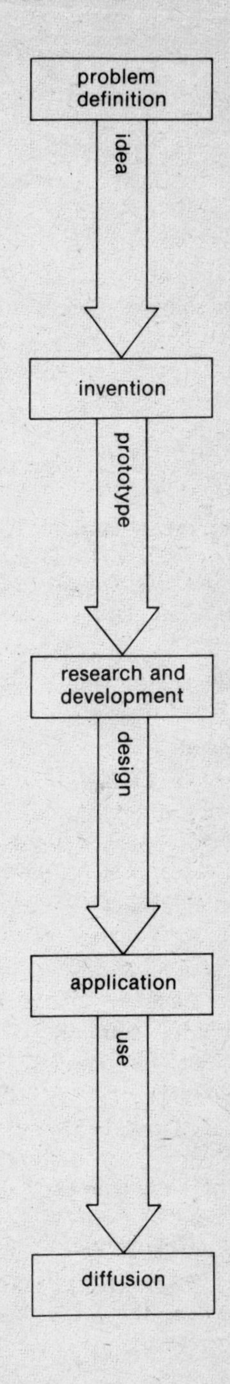

Source: Kelly and Krazberg (1978)

Figure 1 Linear-sequential model

An important point to be made is the relation of design to technological capability; the technology exists and there are a number of ways that design can use this capacity. A crucial determinant is the priority given to particular criteria, that is, cost-reducing criteria, ease of manufacture, user requirements, particular market focus. The design of a product or process reflects particular aims which can give priority to economic considerations, social criteria, design for special needs and so on. The criteria are selected by those involved in the decision-making process, thereby shaping the subsequent technological trajectory, while being influenced by and reinforcing what Dosi (1982) terms the 'technological paradigm'.

Technological Paradigms and Technological Trajectories

The argument that radical innovations form the basis of the development of a trajectory (see Rosenberg 1982) has been developed by Dosi (1982) in terms of technological paradigms and technological trajectories. 'Technological paradigm' refers to a model or outlook that is brought into existence by a 'radical' innovation. For example, advancements in solid-state physics influenced innovations within the electronics industry. The 'technological trajectory' refers to the process of continuous product improvements, for example, computers in the electronics paradigm.

For Dosi (1982, p. 148), the technological paradigm 'is an "outlook", a set of procedures, a definition of the "relevant" problems and of the specific knowledge related to their solution . . . each technological paradigm defines its own concept of "progress" '. The concept of paradigm is based on Kuhn's (1962) scientific paradigms. Within this view, the paradigm includes certain possible technological alternatives and routes to future technological developments. For example, if nuclear technologies are selected as the 'model', then energy problems are defined by nuclear technological solutions. Acceptance of particular technological paradigms can play a powerful prescriptive role with regard to the selection of routes of technological development. That is, solutions to technological problems are provided by specific technologies and not others, for example, to return to the subject of energy, the technological paths defined by nuclear power mean that other energy technologies, such as wind power and solar power, are excluded.

The exclusion effect of technological paradigms means, as Dosi (1982, p. 153) writes, 'that the efforts and the technological imagination of engineers and of the organisations they are in are focused in rather precise

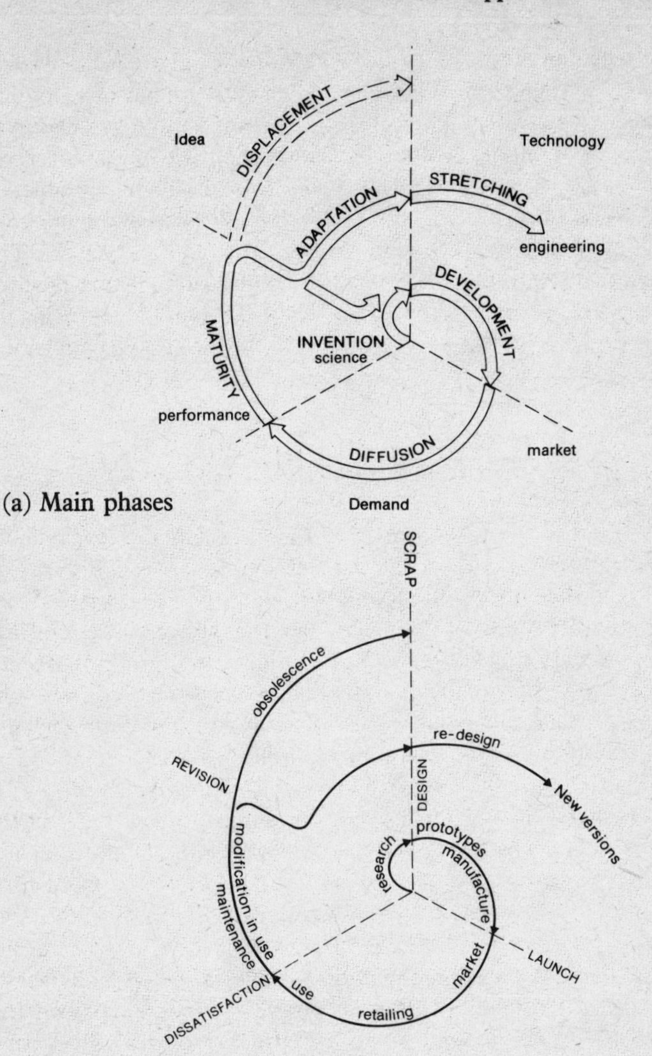

(a) Main phases

(b) Typical sequence for product development. The trajectory can either continue to the scrap heap; step down to critical modifications based upon the existing artifact and infrastructure; or contract to the centre, where the process of invention begins again.

Source: *Design: Processes and Products* (T263), Units 5–7, Milton Keynes, Open University Press.

Figure 2 Innovation spiral.

directions while they are, so to speak, "blind" with respect to other technological possibilities'. Progress is defined in terms of the paradigm.

The questions which this framework of thinking about innovative activity raise are: why certain technologies emerged instead of others; how technological paradigms are established; and which factors act as 'selectors' in the sense of identifying certain paths of development within a larger context of possibilities. Discussion of the innovation process suggests that trajectories of development are influenced by economic factors, market demand and profitability, together with the technological capability of the enterprise. The wider social and political environment influences the selection of technological paradigms that act to institute technological trajectories.

Economic and Institutional Determinism

For all innovations, finance plays a central role. The technological 'know-how' may exist but until investment is considered necessary for continued competitive advantage or potential profitability, technological potential will not be transformed into an actual product or process. A finding of the empirical studies of innovation, mentioned earlier, alluded to the uncertainty involved in all attempts to innovate. However, with 'incremental' innovations the risk is considerably less than with 'radical' innovations. Indeed, as Walker (1986, p. 25) says, 'radical innovations come from an attack on matters of principle . . . incremental innovation and new product designs come from the application of engineering practices to design details'. Moreover, the first route is now more likely to be funded through government agencies, while the second route operates by typical market forces.' Government intervention in technological innovation is more and more commonplace in competitive market economies. There are two main functions for this trend, namely, technological support and regulation. On the one hand, there is an assumption that increasingly sophisticated technology results in higher economic efficiency and competitive advantage in international markets; this is a strong motivation for governments to support, foster and accelerate technological innovation. On the other hand, fears about the potentially undesirable social and environmental impacts of technology force governments to involve themselves in a range of regulatory activities.

For radical innovations, reliance on the market to evaluate a new product is wellnigh impossible because there are no comparable products. Because of this the uncertainty involved in such innovations is relatively high and normal loan finance is difficult to obtain, however risk-orientated. Financial support for research and development has traditionally been the main type of public support policy for technological innovation. Once the development stage has been reached, public support usually ended as it was assumed that marketing, start-up, production and all other activities related to diffusion were commercial activities and so should be supported by private industry.

Priorities for public expenditure on R and D and innovation support are fairly clear-cut in Britain. Defence and areas of potential economic importance such as information technology (Alvey 1982) typically receive much greater levels of support than welfare and environmental research. During the years 1972–83, expenditure on defence related research rose by 22 per cent and the resources allocated to basic research declined by about 20 per cent. In 1985, military related research expenditure accounted for 5.4 per cent of Gross Domestic Product (GDP) in the UK. This compares with 6.6 per cent of GDP in the USA, 4.1 per cent of GDP in France and 3.3 per cent of GDP in West Germany (SIPRI, 1986, p. 243).

Selection based on political and economic decisions has a powerful effect on the direction of technological advance, effectively closing off some options, for example, the decision to invest in nuclear power rather than renewable energy has left the latter marginalised, at least in the UK. Other examples point to the decisions to invest in projects that exhibit technological advances but are not the most urgent in terms of social needs. For example, Freeman argues that this is the case with medical services; whilst heart transplants are an important scientific-technological advance in surgery there was no attempt to relate the costs of development to the costs of other more urgent socio-medical needs (Freeman 1982, p. 303).

In the UK the current concern is to promote high technology in the form of microelectronics, information technology (IT) and biotechnology. This is illustrated by Department of Industry (DoI) and British Technology Group (BTG) support for microelectronics. BTG gives a high priority to investment in firms wishing to innovate in the areas of microelectronics and robotics. The Product and Process development scheme run by the DoI has a budget of £80 million to spend on the support of IT over a four-year period (see Braun 1984). The DoI is concerned to support both the development and applications of manufacturing technologies in the areas of Computer-Aided Design (CAD) and Computer-Aided Manufacture (CAM). Robotics has also received government funding for R and D, and in addition the government has given support to firms introducing robotic technology into their manufacturing processes.

The Ministry of Defence (MoD) has played a key role in the support of the manufacture of electronics components, it finances R and D work and is a large customer for electronic components and devices. The high priority given to defence-related research in the UK has led to concern that research resources are being diverted from other sectors of the economy. Walker (1986, p. 225) argues that 'traditionally, the high cost of the defence research and equipment budget has been defended on the grounds that it has a beneficial spin-off effect into civilian industry'. However, it has been pointed out by Sir Ieuan Maddock, former chief scientist at the DoI, that this may not be the case. In NEDO (1983) he notes: 'What was striking was the distance between the attitudes of the civil and defence oriented companies even when they existed within the same group. There already exists a large culture gap and it is getting even wider' (quoted in Walker 1986, p. 225).

This discussion has sought to highlight the areas that have been deemed economically and ideologically important and which have attracted government support. In Britain high technology is very much the technology of the moment. Moreton (1983) sums this up when he says: 'The country is rushing, helter-skelter, into a new technological world of laser beams, electronic beams, computer hardware and software, microfoils, fibre optic technology and diagnostic reagents.' (Moreton 1983).

The Selection Process

High technology, which serves both an economic function (prospective competitive advantage) and an ideological function (high technology is linked to notions of progressiveness), is a selected area of technological development. The high-technology paradigm effectively excludes support for other industries and sectors that are not considered 'sunrise industries'—textiles, coal, and so on.

In terms of Ray's product and process assessment criteria this would indicate that innovations which fit into and harmonise with other high-technology areas are more likely to be commercially successful, once the high-technology trajectory has been established.

The applications of 'high technology' to social or environmental sectors tend to be constrained by powerful economic and institutional interests, with defence preoccupations often dominating. An editorial in

Electronics Weekly (March 1986, quoted in Elliott 1986, p. 17), states this point in the context of the British electronics industry:

> Electronics in the UK has been largely dominated by defence requirements. The big companies have dominated the queue for defence contracts. Those big companies have then been able to decide which part of a contract's content is to be performed in-house, which part will be subcontracted, and which part will be brought in from outside suppliers.
>
> In consequence, the big companies have been able to determine in which areas of engineering the UK is to operate. All big companies tend to fear three things above all: the growth of new companies which might threaten them; becoming reliant on employees who are not committed company men; and becoming involved in areas of activity they can't control. Consequently, it is in the big companies' interests not to have any UK engineering involvement in areas where new companies can grow quickly—which tend to be resistant to big company loyalty ideals; where there is a requirement for up-front expense combined with uncertain returns.

Further selecting devices for technological paradigms within capitalist economies are what Nelson and Winter (1977) term 'natural trajectories toward mechanisation and exploitation of economies of scale' (cited in Dosi 1982, p. 155). The current application of microelectronics to production processes can be seen to reflect this trajectory.

In sum, powerful economic and political factors act as 'selectors' of technological paradigms. In this way the general direction of technological development is defined—the strong influence of a concept of progress acts to delimit the technological trajectories within the overall paradigm. Once established, the path to development 'shows a momentum of its own' (Dosi 1982, p. 156) within the context of the market.

Patterns of industrial and social conflict can also serve to 'select' particular paths of technological development. Analysis in this area may be fruitful for the discovery of some relationships between the choice of technological paradigm and social factors, for example, Noble's (1979) study of numerically controlled machine tools shows how automation can be chosen over other forms of technological development for reasons of management control. This is discussed more fully in Chapter 2.

The selection of technological paradigms is influenced by political, institutional and economic factors. As a selector the role of the consumer market is weak. The market selects among a range of technologies already chosen on the supply side. While people make their own choice from what is available on the market, the range and nature of what is available is shaped to a greater or lesser extent by the economic interests of those who benefit from the economic exchange process.

The argument that technology does not reflect needs forms the base of much of the criticism of the direction of technological advance referred to in the Introduction. In the next section market 'needs' are discussed in order to identify in what areas technology does not reflect needs.

Market 'Needs'

Our discussion of innovation so far points to the interaction between technology and market needs as motivations for improvements in the design of products and continued innovation. But the concept of 'needs' has yet to be explored, in particular in relation to 'market demand'. The fact that the needs of a variety of sections of the community have not been satisfied or have been badly catered for, for example, in the areas of public health, leads to the conclusion that the existence of needs does not necessarily lead to the production of the means to meet those needs.

The concept of 'needs' is tautological and difficult to classify (social and political needs are discussed in Part Two). In the context of a discussion of the market mechanism, needs referred to are economically-defined needs—needs do not become marketable demands until backed by purchasing power so that it is in the interests of the producers to respond. For example, an individual may have the need for a particular specification of wheelchair—if he/she does not have access to economic resources with which to pay for it then the need will remain unmet.

The market mechanism is presented as the user's and consumer's power to choose and therefore to influence the innovation process. There are a number of factors which severely delimit the consumer's and user's power to choose.

As noted above, access to economic resources is the main constraint on free will for sections of the population. In addition, mass-production techniques cater to majority interests so that the needs of minority interests tend to be more specialised and therefore more expensive. The market does not reflect the diversity of group interests.

In an ideal consumer market, the consumer theoretically has the power to choose between a number of alternatives, and the firm is seen as ultimately subordinate to the market and consequently to the consumer. 'Consumer sovereignty' is the term used to describe this state of affairs; the ideal model presents a picture of the consumer in possession of perfect information and able to choose freely between goods for the best price and quality. The competitive mechanism ensures that suppliers adapt their output to meet these needs. This model can be criticised in a number of ways but, perhaps most importantly in the notion of the competitive mechanism.

The market as characterised by large numbers of small, competing firms has now largely disappeared from most economic textbooks to be replaced by the more realistic interpretation of competition as that which exists between a relatively small number of large corporations. The free enterprise system and the rule of competition between essentially passive firms has been replaced by the notion of monopoly capitalism—the term used by Marx to describe this stage in the historical development of capitalism. Most economists and social scientists would agree that the competitive model no longer exists but would arrive at their conclusions through different methods of analysis. What this description of the market means for the theory of consumer sovereignty is that the producer in effect controls prices and output. The characteristics of the consumer market as oligopolistic and primarily concerned with product differentiation and planned obsolescence led Galbraith (1969) to develop his theory of 'producer sovereignty'. The management of consumer demand through advertising further erodes the power of the consumer.

In terms of technical innovation, Freeman (1982) points to three main ways in which consumer sovereignty is eroded. First, the nature of consumer choice is static. Consumers choose between the existing goods on offer, however in areas of rapid technical change these goods will have been the result of decisions to innovate or apply R and D a number of years earlier—consumers do not have the power to influence the future array of goods. It can be argued that consumers have an indirect influence in the sense that firms are concerned to anticipate demand in order to make a profit. Whilst this may be true, it is also possible to argue that, in fact, those who make the decision to innovate impose their choices above those of the consumer.

Second, the notion of consumer choice also implies the availability of perfect information about the array of goods. The increasing technical sophistication of consumer goods such as cars, televisions, videos, and the whole range of electronic equipment now on offer make it extremely difficult for the consumer to gain access to full and accurate technical information which is understandable to the lay person.

Technologies reflect cultural conditions; goods are linked to beliefs about social status, social acceptability and personal achievement. Perhaps one of the clearest examples of this is the status that attaches to owning a Rolls-Royce car. The market as a restricted and socially-conditioned mechanism cannot reflect all aspects of a community's beliefs about social usefulness. Market production is an indirect means of ensuring the usefulness of a product, commercial success motivates the supplier to make products that will secure economic advantage; the actual usefulness of the products themselves is secondary. Thus, for example, it is the profitability of producing pharmaceutical drugs rather than the curing of illness which is the prime motivation of the drug companies.

This discussion of consumer choice on the market points to the constraints on individuals who do not have the economic power to influence the innovation process from the demand side. Of crucial importance is the inability to influence the future array of goods and services. Although markets act as a selecting device, it is selection within constraint and, as Dosi (1982, p. 156) argues, 'generally among a range products already determined by the broad technology patterns on the supply-side'.

In the capital-goods market, 'needs' reflect economic priorities and the need for efficient production processes for products operating in a competitive environment. In the later phases of a new product innovation the emphasis tends to move away from product improvements towards a concern with cost reduction and productivity gains (Gershuny 1985). The requirements of efficiency act as a constraint on choice of production techniques; these 'needs' of management may conflict with the 'needs' of labour for improved working conditions. The acceptability of new production processes will involve a political compromise between these two groups. In the present state of affairs this compromise will be reached during the stage of implementation of the technology rather than during its design and development. In Chapter Two the possibilities of building the 'needs' of different groups into technologies during their design and development will be discussed.

For many groups in society the ability to influence the innovation process, whether on the supply side or the demand side, is limited. The recognition of this by several local authorities in the UK has prompted the attempt to introduce technology initiatives based on social need and user involvement in the innovation process. These initiatives will be discussed later.

Summary

Economic approaches to the study of the innovation process are concerned to explain the origins and development of the machine or tool by reference to factors which influence innovative activity. The criteria used to define 'successful' innovations are economic indicators of profitability and market share. It has been argued that economic influences, together with ideological factors, in the sense of prestige-type projects, act as an outlook whereby certain patterns of technological development are selected as the route to technological and social progress.

The use of the concept of 'technological paradigm' developed by Dosi (1982), although 'impressionistic' and difficult to analyse empirically, provides an explanatory framework for understanding the nature and direction of innovative activity in competitive market economies.

The discussion of government intervention highlights the role of political processes in the selection of technological areas to support—in Britain this has taken the form of particular sectors of 'high-technology' development. The allocation of resources to R and D in these sectors is based on their perceived economic importance. The identification of technical change with economic growth acts as a powerful argument in support of chosen patterns of development. Technological and economic determinism serve to present the selection process as 'objective' and neutral, not political. Perceived economic importance is seen as the 'objective' reason why certain technologies emerge instead of others.

Technology-push approaches to the explanation of the innovation process are useful for understanding the nature of 'radical' innovations. These innovations advance cumulatively along technological trajectories within overall paradigms of development. Distinctive of developments within the innovation process throughout this century are the growth of specialised knowledge and the institutionalisation of R and D in industrial and government laboratories. The innovation process is increasingly privatised and dominated by 'scientific and technical experts' whose vision is focused on technological problems and their solutions within a defined trajectory and overall paradigm. In this way, 'experts' become blind to alternative technological possibilities. The dependence on economic resources for the development of innovations in an uncertain environment means that in certain sectors government support is a key factor. In the allocation of scientific resources governments influence the overall technological environment. Their role is central to the selection of technological paradigms. The priority for defence-related technologies (for example, the development of electronics) is evident in Britain, where government procurement also plays an important role.

The privatisation of the innovation process effectively excludes a wide range of social groups, who do not have access to political or economic power, from participation in the decision-making process on the patterns of technological development. Decision-making for these groups is confined to the democratic electoral process once every five years!

Market-pull explanations of the innovation process are useful for understanding 'incremental' innovations. Radical innovations only account for a relatively small part of innovative activity; most innovations tend to proceed through a series of major and minor improvements. Theoretically, the influence on incremental innovation is the recognition of need; however, the satisfaction of need through the market mechanism is the satisfaction of economically defined needs. Within the consumer-goods market there is a selection among a range of products already determined by broad technology patterns chosen on the supply side. Consumers do not have the facility to influence the future array of goods, rather they choose between the existing array of goods on offer. In the capital-goods market the dominant 'needs' of efficiency and reduction of production costs are reflected in automated manufacturing processes. The ability of labour to influence the technology is *ex post facto*; influence is possible in the process of implementation and work organisation but the technology itself is already designed and developed—thus influence is within the constraints of the hardware.

Models of the innovation process are essentially prescriptive and designed to facilitate 'more' innovation. The breakdown of these models of innovation management into various phases serves to aid the understanding of the factors and processes involved in the development of technological artefacts. In addition, models are useful as analytic tools—they make it possible to identify social groups involved in each phase of the innovation process. In the present state of affairs the criteria that are used for solving the technological and other problems within phases of the innovation process are focused on certain technological trajectories where economic factors take priority.

Conclusions

The list of unmet 'social needs' and problems for which technological solutions could be possible is endless. In the 'advanced' world there are many who still die of renal failure, yet heart-transplant surgery is well developed. In the developing world, thousands starve while the West spends millions on developing weapons systems. It is possible to put astronauts on the moon but the provision of movement aids for the handicapped is rudimentary.

These examples indicate that the problems are not just technological; the influence of a variety of social, economic and political factors affect the development and deployment of technologies. The existing economic and political structures and criteria dictate that certain groups have access to the necessary financial, technical and material resources to survive and prosper, whilst others do not. The contradictions between what technology can provide and what technology actually does provide stem from an unequal distribution of resources among social groups; what technology actually does provide reflects the interests of groups in society who have access to economic and political power.

Why certain technologies emerge instead of others, and whose needs are reflected by the market, are pertinent questions that relate to the social process of innovation and the way it is organised. The conception of technology as machinery removed from its social context of implementation is embedded within economic analyses of the innovation process. Yet much innovative activity is dependent upon knowledge gained from its social implementation. The innovation process and the social implementation of technology are fundamentally the same process. Wynne (1983, p. 16) writes: 'The social organisational problems of enactment often dwarf the supposedly key problems in innovation—namely the complex technical problems usually addressed by R and D. Their neglect by design/innovation actors undermines the viability of technology.' A prime example is the computer-software industry, where effective software is extremely dependent on user experience. Rosenberg (1982) points to the importance of 'learning by doing'; for many innovative improvements, innovation is followed by a lengthy process of redesign and readaptation. Of key importance are the social groups involved in the design and development of new products.

Investigation of the innovation process that recognises its close relationship to social implementation requires a social conception of technology. If the innovation process is seen as reflective of the interests of certain dominant groups in society, for example, decision-makers, technical and financial elites, then the involvement of a different range of social groups will affect the direction of the innovation process and technological development. The social conception of technology makes explicit the 'multi-directional' nature of the design/innovation process. There are always alternative designs of products and processes and alternative solutions to technological problems. The creation of facilities for the social groups excluded from privatised innovation processes is a political and organisational problem which has been addressed by the 'radical' local authorities in their attempt to redirect resources to meet social needs through the development of socially-useful technologies.

In this chapter the emphasis has been on the factors that affect the direction of technological change; there has been little discussion of the technology itself. Chapter 2 looks at the way that political and social factors affect the design of technology itself.

Notes

1. Queen's Award to Industry for Economic and Technical Success.

2 The Social Process of Innovation

In Chapter 1 it was argued that powerful political and economic factors influence the selection of routes for technological development so that technology is neither neutral nor autonomous, rather it reflects particular policy objectives and societal goals. This point is not new—over a decade ago, Brooks *et al.* (1974, p. 140) argued that: 'The capability of governments to introduce new socially-oriented technologies, altering traditional lines of technological development, is often less restricted by objective factors, such as competitiveness on international markets than is commonly assumed.' They also make the following prescriptive point in relation to their concern for the development of socially and environmentally appropriate technologies:

> The key requirement is a wider range of options in the early stage of the innovation process combined with a more sensitive, comprehensive, and rigorous process of choice as the various options progress towards application. This involves deeper consideration and exploration of alternatives at the beginning, with a larger number of checkpoints in the process of selecting options so that vested interests, sunk costs, and professional commitments do not build up a momentum that becomes difficult to reverse (Brooks *et al.* 1974, p. 139).

This implies that the innovation process is flexible and that there are alternative paths of development. However, do products and processes emerge in particular forms, rather than others, primarily because of general political and economic choice, or are there additional and more specific reasons why products are designed and developed in a certain way? In this chapter the aim is to explore the social construction of technological artefacts and the implications of this for the innovation process.

The innovation process has been identified as a key point of intervention by those concerned to change the direction of technological advance away from an emphasis on the quantity of goods and services and exchange-value toward use-value and the quality of goods and services in their reflection of social need. Social factors affect the shape of technological development during the R and D and innovation stage, and during the use, diffusion and applications stage. Although these stages overlap—much 'innovation' occurs in diffusion—these points offer the possibility of influencing the direction of technological development.

It is perhaps helpful to reiterate the main points of the 'social construction of technology' perspective mentioned in the Introduction.

1. Technology is socially constructed and technological artefacts are influenced by the social relations of their creation, development and implementation.
2. The social conception of technology, by making links between the innovation process and the social implementation of technology, gives a greater understanding of the nature and wider social context of the innovation process.
3. The innovation process is multi-directional, that is, there are always alternative solutions to technological problems.
4. The design of alternative products is based upon the inclusion of a different range of social groups and forces in the social process which shapes technology.

In this chapter we develop these points further, and look at the concepts of 'social need' and 'social use'.

In the social conception of technology perspective the definition of technology is widened to include the social context within which machinery is developed and used, and the overall system of knowledge that supports technological systems. This wider definition clearly illustrates the relationship between technology and society as interactive. Machines or tools are designed and manufactured by people in a social context, the use of machinery by people gives technology its reason to be. For example, what is a car without a driver, or a production line without an operator? Technology is also knowledge (see Layton 1974) in the form of the expertise needed for the design, use and maintenance of physical artefacts.

Once it has been established that technology is dependent on a social context for its development and implementation, then it becomes possible to ask how the social context affects technology; what sort of influences do social relationships have on the design, development and use of technologies?

The Social Construction of Technology

The social conception of technology recognises the interdependence between technology and society in a social context. Within this perspective the social groups involved in decision-making about technologies and how this affects the innovation of new products and processes are seen as important. The point is to attempt to reveal alternative solutions and then explain why certain 'solutions' were chosen rather than others. If it is possible to demonstrate that a technological artefact has alternative possible routes of development that are equally viable then the belief that there is an immanent logic of technical development is questionable. The argument that there are possible alternative solutions to our technological problems is strengthened by the ability to offer alternative solutions to problems/issues that are not on the agenda of mainstream technological development. The claim that there are purely technical reasons for the rejection of certain alternative technologies is shown to be false. The design of technologies is seen to be influenced by social choices, thus the opportunity exists to choose alternative paths of technological development.

As noted earlier, the theoretical basis of the social construction of technology involves two approaches. The first has grown out of the concern of the sociology of scientific knowledge to explain the definition and legitimacy of scientific claims—questions concerning the epistemological status of scientific knowledge. The second approach is in the Marxist tradition.

Studies of the processes of construction of scientific knowledge in the context of the laboratory and scientific controversies (for an introduction to the literature see Shapin 1982), have led to the conclusion that scientific knowledge may be defined as one of a number of knowledge cultures. The acceptability of scientific claims depends upon the social context of their interpretation rather than particular technical or logical criteria. From this perspective, scientists are seen as primarily 'social actors' who use their cultural assumptions, intellectual assumptions and location within a social context to interpret the validity of knowledge claims. The main point is that it can be shown that 'scientific facts' can differ between groups of scientists working in various social environments and with different methodological perspectives. Scientific research is not an 'objective' process of demystification of the natural world, rather it involves the attribution of meaning to physical phenomena within different social contexts. The physical world may act as a constraint but it does not determine the outcome of research and scientific conclusions because it is necessary to understand the influence of the social and political world that affects the interpretation of the meaning of physical phenomena. (For discussion see Collins 1983; Knorr-Cetina and Mulkay 1982).

For the purposes of our argument, the importance of the findings of the sociology of scientific knowledge rest largely on the view that if scientific facts are socially constructed then it is possible to argue that the technical configurations of physical artefacts are likewise open to interpretation and negotiation. If followed through, this would mean that the way physical artefacts come to be designed and developed represents one route amongst a number of possible alternatives.

The logical extension of the belief that science is socially constructed to incorporate the idea that technology is also socially constructed is over-simplistic in that it is based on the notion that technology is the application of science. The science–technology relationship discussed earlier in terms of 'technology-push' emphasised the inadequacy of this approach to technology. Historical studies of technology have clearly illustrated that technology as applied science (which implies a unilinear path of technological development) is a misnomer. Science can influence technology in the same way as previous advances in technology can affect and influence further technological developments. Langrish *et al.* (1972) make the important point that when science does contribute to innovation, it is in the form of personnel trained in the techniques of investigation and in the form, though rarely, of the establishment of a direction for further technological development.

The science-technology relationship is found to be interactive. If science and technology are interactive, then it becomes possible to suggest that political, economic and social factors interact in the same way within the construction of science and the process of technological innovation (see Hughes 1986).

Callon (1980), in a case study of the post-World War II effort of the French state to promote an electric vehicle, uses the concept of 'actors' interacting within a network to subsume the separate categories of science, technology, the social, the political, and so on. The 'actors' include the diverse groups of people who influence the development of the technology: technologists, scientists, users, manufacturers, government departments, and so on. There are no rigid distinctions between the social, political and technological; rather, what is important is the assignment of roles within a 'scenario' or programme developed by Electricité de France to fulfil its objective of designing an electric vehicle. Individuals and organisations are seen to need to interact harmoniously as a precondition for the innovation of the physical artefact. This case study points to the way that the technological meaning of a physical artefact is socially constructed, that is, the technology embodies the social and political environment of its conception, design and development. Design considerations include possible future government regulations, market orientation and user expectations—these do not form the background of innovation—rather technologies embody these influences. For example, user needs for more luggage space and easier loading are embodied in the design of the hatchback car.

Callon argues that the 'actor-world' scenario necessary for effective innovation is not constant but changes and at times breaks down. The insights into the social system of innovation provided by Callon's study lend support to the notion of technological paradigm discussed previously. This notion has a bearing on government intervention in the innovation process in which the overall framework or technological environment is affected by government programmes to support certain technological areas. Callon's point about the changing nature of the 'actor' world in the interaction with technological artefacts allows the possibility of seeing paradigms as offering a framework for 'actors' involved in innovation that shifts depending on the context of the attempted innovation. The technological paradigm is not fixed but acts as a 'resource' (MacKenzie and Wajcman 1985, p. 11) that can be used. Technological paradigms and trajectories and selected routes of technological development within a particular paradigm are neither constant nor determined but are social constructions closely allied to the social context in which innovations are attempted.

The concentration on the 'actors' involved in the innovation process clearly reveals the value of their effect on the nature of technological artefacts. The historical analysis of technological innovations proves also to be a valuable method for the identification of social processes which act to construct the technical meaning of physical artefacts. The identification of alternative solutions to the design of technological artefacts allows the possibility of uncovering both 'successes' and 'failures' so that the question as to why certain designs are selected and not others can be addressed.

Pinch and Bijker (1984) describe the developmental process for technological artefacts as evolutionary or as a pattern of variation and selection. (This is supported by Ray's (1985) characterisation of the innovation process as evolutionary.) The attempt is made to present a multidirectional model of the innovation process that demonstrates that the 'successful stages in the development of an artefact are not the only possible ones' (Pinch and Bijker 1984, p. 411). In their study of the development of the bicycle they show how the penny-farthing was rivalled by a range of possible variations, a fact that has been obscured in linear-sequential explanations. Bruce (1984) demonstrates the same point with reference to videotex. The question which arises is why, if innovation at this point was multi-directional, did some forms of bicycle survive whilst others did not? How did this process of selection operate? To answer this question, the 'social groups' involved in the development of the bicycle are seen as crucial. Pinch and Bijker (1984, p. 414) state that 'a problem is only defined as such, when there is a social group for which it constitutes a "problem" '.

The concept of social group is used to define institutions and organisations (the military, for example) as well as organised or unorganised individuals. The key factor is the shared set of meanings attached to the specific artefact by members of the group. The identification of different social groups involved with the design and development process shows how each group's interpretations of the problems and possible solutions differ. The ensuing conflicts act to explain why the technology emerged in the way that it did. For example, differing technical requirements for speed and safety may be attributed to pressures from opposing groups which affect the final solution or design. The developmental process is seen as consisting of progressive degrees of 'stabilisation' over a period of years. For example, until the safety bicycle came to be seen as the way to design a bicycle, there was a time-lapse of nineteen years during which time other forms of bi- and tricycle competed for selection.

Within the perspective of the social construction of technology, the main point is to reveal and explain how the technical make-up of a physical artefact emerges as the dominant design, that is, how one specific form is selected from the various designs present within the design and development process. As with the sociology of scientific knowledge, where scientific 'facts' are found to be negotiable, technical 'facts' are also seen as negotiable.

The concentration on the social processes of innovation reflects interest in the way that 'actors' or 'social groups' bring their own interpretations to a specific problem-solution. However, explanations in these terms separate the innovation process from implementation of the technology—a product or process is designed and developed and then implemented in the wider social context. The 'social groups' in Pinch and Bijker's analysis who have a new input into the innovation process are those who hold some sort of resource; interest groups who have technical, political or economic resources. However, there are many groups in society—women and ethnic minority groups, for example—who do not have technical expertise or political or economic power. Further, the problem of social choice exerted by powerful interest groups is not discussed. Selection implies a consensus, but a consensus among whom?

Who decides which variations of artefact constitute the choice in the first instance? How is progress or success defined? (For further discussion, see Russell 1986.) The differing abilities of social groups to influence the design, development and adoption of technologies requires explanation.

The social construction of technology approach concentrates on the design and development of 'hardware' as the embodiment of the social context of innovation. The central concern is to dismiss claims of technological determinism by illustrating the existence of social factors which affect the choice between alternative paths of development. The emphasis is on 'workability', 'how engineers and technologists actually go about deciding whether or not a technology works and how it is to be tested' (Pinch and Bijker 1986, p. 351). The choice over technological options and the complexity of other societal factors in the shaping of technology point to the possibility of alternative paths of technological development. However, if the innovation process is extended to include the implementation phase where technologies are adapted and readapted for use, then it becomes possible to argue that investigation of the process of social choice, by those with the power to choose, offers a clearer way to understanding of the societal influence on technology.

Langdon Winner, in an article entitled 'Do Artefacts Have Politics' (reprinted in McKenzie and Wajcman 1985), discusses the design of technological artefacts as exhibiting conscious social choice for desired effects. He cites the example of the building works of Robert Moses with particular reference to the bridges over the parkways on Long Island, New York. The bridges were built to specifications that precluded the presence of buses on the parkways; an effect of this was to limit the access of racial minorities and low-income groups to Jones Beach public park. The bridges embody Moses' social class bias and racial prejudice in that car-owning white people of the 'upper' and 'middle' classes are free to use the parkways for recreation and commuting. Winner argues that technology is inherently political; it is designed consciously or unconsciously to open certain social options and close others.

In order to understand the reasons for the choice of certain forms of technology as opposed to others it is thus necessary to locate technology within the social context of its use: how it operates in the real world. The influence of 'social groups' on the design and development of technological artefacts requires an understanding of the unequal distribution of resources and power in society.

The Marxist Approach

The Marxist approach to the study of technology points to the interaction between the economy, society and technology. An understanding of the dialectical process for the formation of products and processes is important. It forms the base for the emphasis on social processes of innovation and technological development as the appropriate area of study. Technical change cannot be understood by reference to individual inventors or particular innovations, rather it is necessary to examine how larger social forces affect the focus of technological problems that require solutions. Usher perceives this as the question: 'how is the stage set to suggest the solution of the perceived problem?' (cited in Rosenberg 1982, p. 49).

The Marxist view of the analysis of science and technology has largely concentrated on the labour process. In particular, the nature of control is emphasised. Much of the research on labour process has grown out of the analysis of the nature of technology in capitalist society by Braverman (1974). The central concepts in Braverman's analysis are 'deskilling' and 'taylorisation'; this refers to the use of management principles, firstly, to dissociate the labour process from the skills of workers; secondly, to separate conception from execution; and thirdly, to use the resulting monopoly of knowledge to control each step of the labour process and its execution. In this analysis, the process of taylorisation is not a result of technological change within the production process, the technology itself is seen as neutral; taylorisation represents a form of implementation of technology that ensures capitalist control. The extensive research which has followed Braverman's analysis has shown that 'deskilling' and 'homogenisation' are a tendency within capitalist technological implementation (see Wood 1982) but that implementation is affected by other factors apart from capitalist control.

Several studies have shown that worker resistance is an important factor in the technological transformation of the labour process (Burawoy 1978; Elbaum *et al.* 1979); further, that the relationship between technology and skill requirements is affected by local conditions and wider economic and political factors (Wood 1982). However, the concept of the straightforward use of technology by capitalists to control workers is oversimplistic and ignores other mediating factors which influence innovation and technological change. Elbaum *et al.* (1979) point to the way competition acts as a constraint on the capitalist use of technology for the control of workers. In addition, the choice for managers to adopt new technologies from the available alternatives on offer involves a process of political negotiation between management and workers. Wilkinson

(1983) argues that technological choice and the establishment of working practices are negotiable rather than the outcome of a system beyond the control of actors and interest groups. The mode of implementation of technologies and subsequent 'debugging' processes offer opportunities for actors to influence the process of technical change and innovation in the diffusion phase.

From this perspective, the ability of groups of workers to influence technology is *post hoc*. What is unclear in the labour-process analysis of technology is the issue of the actual shaping of technology—are social relations embedded in technology or do social relations affect the application of technology? This question as to whether the design and development of machinery reflects the social relations of its conception is left wide open.

Other research in the Marxist tradition has addressed this question. In a study of numerically controlled (NC) machines tools, Noble (1979) argues that social relations are embedded in technologies. This study of the selection of technological variations by those with the 'power to choose' makes the point that technologies of production reflect the social relations of production both in design and implementation and in the realities of shop-floor struggles between classes.

The aim of the study is to discover why the technology, NC machines, took the form it did. The 'problem' of automating contour machining offered two solutions: 'record playback' and NC. According to Noble, the difference between these two solutions rests on the amount of control over workers' skill that was offered by the technologies. Record playback involved obtaining repeatability whilst the intelligence of production was left in the hands of the machinist who made the tape. NC, in contrast, reflected the manufacturing philosophy of the breakdown of machining into numerous separate tasks so that the intelligence of production was built into the machinery, replacing the skill of the machinist. Noble argues that the reasons why NC was developed and record playback was not, can be attributed, in large part, to the issue of control. During its development NC received substantial military support and its cost meant that only large companies, often in receipt of government contracts, were able to adopt the technology. The ideological factor, NC as 'symbol of the computer age', appears to have been more important than economic considerations. Record playback was suitable for most machining needs, whilst not involving the huge investment in computing and programming necessary for NC.

The social relations of production appear to have played a key role in decisions to develop NC. Noble gives the example of General Electric;

although one of the first companies to adopt record playback, after a series of strikes during the 1940s, machining was automated using NC. The reasons for the shift in design effort from record playback to NC are evident from this quote from an engineer involved:

> Look, with record playback, the control of the machine remains with the machinist—control of feeds, speeds, number of cuts, output: with NC there is a shift of control to management. Management is no longer dependent upon the operator and can thus optimize the use of their machines. With NC, control over the process is placed firmly in the hands of management—and why shouldn't we have it? (Noble 1979, p. 34).

However, as with the studies of the labour process, the issue of control is not the whole story; despite these attempts at control management still needed experienced machinists to ensure that 'automatic' machines worked smoothly and produced components of good quality.

Noble's research illustrates that automation did not have to proceed in the way that it did and that choices other than the purely technical or economic affect the design and development of technologies. These findings are supported by other studies which demonstrate that the social forces affecting technological development are complex and interactive amongst social classes, gender groups, economic and political factors.

Lazonick's (1979) account of the development of the self-acting mule in the spinning industry shows how adult mule-minders in Britain retained their positions of power not because of resistance but because employers found the hierarchical division of the work-force between minders and piecers useful. In this way differential relations among the work-force affected technical development.

In a study of the American garment industry, Schwartz-Cowan (1979) argued that the sewing process had not been automated because of the cheapness and availability of immigrant women workers. In this way gender and class relations affect technical change. This point is reinforced by Cockburn (1981) in a study of typesetting technology in Britain. She shows that the choice between alternative designs of mechanised typesetting was based on the preservation of male domination in the composing room. This is indicated by a statement made by the male-dominated union, the London Society of Compositors: (by not splitting up the typesetting process) 'the Linotype answers to one of the essential conditions of trade unionism, in that it does not depend for its success on the employment of boy or girl labour' (Cockburn 1981, p. 46).

The effect of capitalist relations on technologies is the subject of a study of domestic technology (Schwartz-Cowan 1983). The domination of electric refrigerators over gas refrigerators is shown to result not from technical superiority or consumer preferences, but differences in expected economic returns.

This empirical work suggests that the influence of social factors on the shaping and choices of routes of technological development are embedded in technologies themselves. The identification of alternative possible routes of development that were neglected for social, political or economic reasons highlights the way the nature of technology is inextricably linked to its social context. The key point is that the understanding of the effects of social relations on technologies requires not only the investigation of the design and development of physical artefacts but also their role in the social context of implementation. It is not possible to argue that technology is neutral and its application reflects social relations; rather the evidence of alternative paths of technical development points to the importance of selection and choice of technical design by social groups with the economic and political power to choose.

Implications for the Innovation Process

The view of technology as a social system has implications for the workability of technologies. The labour-process theorists point to the way that workers mediate and affect the implementation of technologies. Within the present social organisation of the innovation process this influence is typically *post hoc*, that is, after technologies have already been designed and developed.

In Chapter 1 the importance of the interaction between science, technology and the market was discussed in terms of prerequisites for 'successful' innovations. The relationship between innovations and users is important. In the economic approach to the study of innovations the saleability and therefore 'success' of technological innovations are dependent on how reflective they are of the needs of the user. However, the current social organisation of the innovation process means that the needs of the user do not have a direct input into the innovation process—rather needs are fed back to producers in the form of market demands. The capital-goods sector operates on a closer relationship between producers and users, where purchasers have a more direct input into design specifications. However, the workability of technological innovations is clearly dependent on the social context of implementation—technology works because people operate it, maintain it and so on.

The view that technology does not reflect 'needs' is based on the evidence of the appropriateness or otherwise of technologies in certain social contexts, for example, lead in petrol is not viable in the context of the social costs of pollution. The concentration on technological innovations as physical artefacts serves to obscure their dependence on the social context. As mentioned earlier Wynne (1983) argues that the 'machine' approach to the understanding of innovation splits technology from implementation so that products and processes are designed and developed within the confines of the innovation process. Implementation is an external factor. One effect of this is that the innovation process is privatised and dominated by 'technical experts' involved in their own goals and priorities within the paradigms of certain problem formulations and solutions. The complex realities of the implementation stage of the innovation process are neglected by 'experts' when choice are made between alternative designs and strategies. The privatisation and increasing specialisation of the innovation process means that 'experts' (usually male) are increasingly socially isolated from the realities of the social system of technology—they operate with technical or machine conceptions of technology.

Empirical research demonstrates the problems that the social isolation of 'experts' from users and the split between innovation and implementation can give rise to. In the area of computing it is noted:

> Systems can be ineffective when they are not well understood by people who use them, provide inaccurate data, demand unusual precision and attention, or are difficult to modify when the kind of information users want changes. These difficulties can undermine the utility of the computer system even when its users are relatively homogeneous and welcome computational assistance (Kling 1980, p. 88).

One major difficulty is the production of computer software that meets user needs.

Computing-systems design concentrates on the technical aspects and coding, the user interface is usually the last part to be designed. In a study of computer specialists, Weizenbaum (1976) notes that although it may appear that they are open to influence by user needs of design, often 'specialists' have rather low opinions of users and tend to use their own perceptions of user needs when designing and developing systems. For many groups of workers, user input into computer software packages is marginal. For example, administrative workers and librarians complained of this at a conference workshop on human-centred technology held at Aston University in 1986.

These examples point to the necessity for designers and innovators to understand the social realities of implementation. The area of computing provides a contemporary example but there are many others. The lack of involvement by users in design and innovation means that often technologies reflect 'needs' as perceived by 'experts'; rather than the 'needs' identified by the user. There are many examples (see Lawless 1977) where the viability of a particular innovation may require resources or expertise or particular cultural patterns that do not exist. This is apparent in many cases of technology transfer from the West to underdeveloped countries. There is an argument here for the encouragement of indigenous R and D based in the local culture and economy and able to be realistic to the resources, possibilities and needs of the local population. The social context of implementation of technology is an essential part of the innovation process.

In Chapter 1 the innovation spiral pointed to the way that readaptation is an integral part of the innovation process, that is, incremental innovations of products and processes are dependent on feedback from information gained from implementation and use of the technology. If technological innovation is seen as a continual learning process then 'learning by using' (the knowledge gained from the utilisation of the product by the final user, see Rosenberg 1982) is an equally important component of the innovation process as the learning involved in research, design and development or producing new scientific knowledge.

In a study of certain capital-goods sectors product differentiation may be driven by learning by using; often the users themselves making important modifications that will be part of the specifications of later models. This was the case in the area of clinical analysers (Von Hippel and Finkelstein 1978). It has been shown that models which have proved more amenable to user modifications have ultimately been more successful commercially. In this way, learning by using creates new information that feeds into the design process of new product development, eventually resulting in modification to the hardware.

Design flexibility or the production of hardware that can cope with future changes and modifications and accommodate learning by using appears to be characteristic of high-technology industries where systemic complexity is an integral part of the product or process (Rosenberg 1982). Computing and software design provide a good example where companies are reliant on user modification in order to achieve the optimal design characteristics of software packages.

The recognition of the importance of learning by using is acknowledgement of the importance of the social system of technology for the innovation process. However the capital-goods and consumer-goods markets are qualitatively different. The capital-goods market is more homogeneous and lends itself to 'learning by using'. The consumer-goods market is heterogeneous and highly competitive, so that 'learning by using' is less likely. It is in this area that there is a need for design and development processes to be more open to interaction with users. If the innovation process is seen as a continuing cycle of design, development and implementation, participation by users may be acknowledged as an important factor in the continual feedback of knowledge that significantly affects the design and development stage of product and process development. The capital-goods market can exhibit a close interaction between supplier and user, while the nature of the consumer market may make this more difficult. The alternative initiatives for innovation discussed in Part Two emphasise the necessity of a close relationship between producers and users.

Socially-directed Innovation

A socially-directed organisation of the innovation process recognises the importance of the social context of implementation. The access of social groups to the design and development stage of technological artefacts is an important prerequisite to socially-directed innovation. This involves the acknowledgement of the importance of all forms of knowledge acquired through the use of the technology. The effect of the close interaction between design and use would result in the workability of technologies in the social world (Papanek 1974). An important emphasis is on flexibility in design to accommodate changing user requirements. What is required is a substantial change in the perspectives and attitudes towards participation in the innovation process. This is primarily a political problem; substantial resources are necessary if the route to technological innovation is to be based in the social context of the user.

During the 1970s, participation in the design process, particularly in the area of industrial design, emerged as a theme centred around the idea of 'design for need'. The focus was on product design with user needs firmly in mind, and in some cases direct user participation in design was attempted, particularly in design areas centred around disadvantaged groups such as the disabled. These attempts at public participation were basically concerned with trying to influence the design activities of 'experts'—planners designers and other professionals.

A very different approach emerged when, in 1976, a group of skilled aerospace workers at Lucas attempted to devise their own plans and designs. The Alternative Plan they developed for the manufacture of 'socially-useful products' grew out of the industrial democracy debate in the UK during the 1970s, but went well beyond calls for participation, with the idea of 'production for need' (Wainwright and Elliott 1983).

The Lucas Plan

In 1976 shop stewards at the Lucas Aerospace Division of Lucas Industries published a major report, an *Alternative Corporate Plan*, outlining a range of non-military products which they argued could be produced at Lucas. The aim was to diversify into the production of products that were socially needed and which at the same time would provide secure employment. Alternative 'socially-useful' products were seen as an alternative to unemployment. At that time, Lucas Industries were attempting to make significant reductions in the Aerospace work-force (it had already been cut from 18,000 in 1970 to 13,000 in 1977).

The Lucas Plan represented a unique attempt at technological innovation in that trade unionists attempted to link the skill and equipment resources at Lucas to social needs which they identified in the community. In some cases prototypes were developed by sympathetic researchers in polytechnics and colleges and several products have been subsequently developed, however now by Lucas Industries. The Lucas Aerospace management were obstructive and hostile to the Plan, possibly because it questioned their right to determine what should be produced at Lucas Aerospace and because it was developed by trade unionists. The result was that after several years of often bitter struggle, the Plan was marginalised, and several key shop stewards were sacked.

Several of the sacked shop stewards subsequently found employment with the 'radical' local authorities in Britain who were trying to develop a similar idea of production for need. Mike Cooley, a leading member of the Lucas Aerospace Combine Shop Stewards' Committee, joined the Greater London Council (GLC) Economic Policy Group and became Director of the Greater London Enterprise Board's (GLEB) Technology Division. Thus in many ways the Lucas workers' initiative was the progenitor of the sort of approach to innovation being developed by the GLC and other councils.

The Lucas Plan proposed 150 products in all, mainly in the areas of energy, transport and medicine, including gas-fired heat pumps, a hybrid road/rail bus and improved kidney machines. Prospective customers were local and national government transport, housing and health departments (replacing the Department of Defence) so it followed that the 'radical' local authorities would be most likely to take an interest.

What are the implications of the Lucas Plan initiative in relation to innovation? In that the Plan was not implemented at Lucas Aerospace it could be seen as a range of failed innovations, even if many of the product ideas were developed elsewhere with commercial success. Although this may have happened anyway the Lucas Plan did put ideas on the technological agenda more formally, particularly those relating to alternative energy systems. At the time the whole renewable energy field was somewhat marginal: the Lucas Plan acted as a ground-breaker.

In terms of the innovation process itself, it is interesting that it fell to a group of workers to develop proposals—drawing on their own technical experience and their knowledge of community needs. It was evident that there was a considerable reservoir of untapped talent and ideas in the work-force, which was unleashed to produce a unique Plan, with the political muscle of the Combine Shop Stewards' Committee behind it. In the event however, the Combine was unable to push the ideas through—so that at the very least its role as a 'product champion' must be queried. But the concept of worker involvement in product development has certainly been extended well beyond the 'suggestion scheme' ideas. Clearly the Plan was developed and presented (and for that matter received) in a conflict industrial relations context. The Combine saw it as a collective bargaining demand, as something to be negotiated with management, in the same way as wages or working conditions. It was not an attempt at joint problem-solving: the Combine saw the workers' interests as fundamentally different from those of top management and the shareholders.

Subsequently, several other groups of workers have developed similar plans—for example, at Vickers in Newcastle and Barrow, at C.A. Parsons, and Clarke Chapman in Newcastle, and at Chrysler. In each case workers faced with redundancy following falling orders for the company's existing products have proposed and campaigned for alternatives. By doing this workers have tried to influence corporate technology policy and product market strategy. The Parsons stewards, for example, have been one of the leading champions of the combined heat and power concept—under the slogan 'Jobs from Warmth'—a slogan later adopted by the GLC. (For further discussion, see Wainwright and Elliott 1983.)

The point of alternative plans is to present constructive alternatives to existing technologies, thereby creating or maintaining employment. For the 'radical' local authorities, the notion of alternative plans and production for social need are key components of their economic thinking (Bennington 1986).

Socially-directed Innovation Policy

For the 'radical' local authorities, the innovation process is a point of intervention for initiatives designed to link innovation directly to need, and thereby create employment. 'Socially-useful production' is the term used to refer to this form of innovation strategy. Cooley (1984, p. 51) defines socially-useful products as

> products that in their design, manufacture and use enhance human skill and ingenuity, conserve energy and raw materials and aid human beings rather than control, de-skill or maim them. In addition, new forms of technology are supported which would provide for human-enhancing and liberating means of making socially-useful products—human-centred technologies.

Socially-directed innovation is an overtly political programme; it has been identified by the 'radical' local authorities as central for economic regeneration and the furtherance of social and political objectives.

A major component of this innovation strategy is participation in the product design and development process by work-place and community groups. The creation of facilities to support technology transfer from higher education institutions to local enterprises in the form of Technology Networks' which are described later served as a back-up to this. If the innovation process is seen to involve, for simple explanatory purposes, the phases of idea-generation, product prototype, development and the market then it is possible to show the inputs and social organisation diagrammatically:

IDEA	community groups
	workers plans
	academic links
	articulated needs
	technical expertise
PRODUCT PROTOTYPE	technology networks
	worker input
	academic links
	participant input
	GLEB finance
DEVELOPMENT	co-operatives
	GLEB supported enterprises
MARKET	public sector agencies
	commercial companies

The Lucas Plan demonstrated the possibility of an alternative social organisation of the innovation process by linking the skill, equipment and resources at Lucas to social needs identified in the community. A missing ingredient was markets and the finance to follow through with the development of products. However, socially useful products in the form of medical technologies, transport technologies, aids for the disabled, and so on, are directly relevant to local authority statutory service provision. The GLC were attempting to institute a form of public enterpreneurship whereby public funds, allocated through the GLEB, were used to meet needs expressed by the community. The idea was to augment the conventional consumer-market mechanism. Socially-useful production implies a contrast to production for profit: 'products' consumed were often not conventional consumer items sold on the open market to individual consumers but rather basic services such as heating systems in council flats provided by the public sector. In this way it was seen as possible to link the innovation process directly to social need.

Needs

In Chapter 1 the inadequacy of the market mechanism to cater for all 'needs' was discussed. Needs in the context of the market are economically significant 'market demands'. Needs relating to collective service provision are dependent on political priorities and the commitment of resources, for example, there is a need for increased provision of public sector housing but not the commitment of financial resources from central government.

Minority-interest needs may be defined as the needs of groups who do not have access to political and economic resources and power with which to articulate and satisfy their needs. Who defines needs, whose needs are more important than others are political questions. The possibility of differentiating between needs and wants, and the recognition of the cultural relativity of needs serve to highlight the problems of basing an innovation process on the satisfaction of needs. Human need is a complex concept which is impossible to pin down theoretically (Doyal and Gough 1984).

In practical terms it is recognised that there are basic needs for food, clothing, shelter, and so on, and that these needs can be met in many different ways, for example, food can be home-grown, bought in a supermarket, or consumed in a restaurant. Choice on the market is between existing alternatives and access to purchasing resources. In addition, there is an area of public provision where needs are met collectively. In a market economy, service provision provides goods that are socially vital but economically inaccessible to large groups of potential users, such as health care, education and social services. In addition, there are needs relevant to particular groups that have not received priority on the technological agenda for political, social and cultural reasons. These include the particular needs of women, of ethnic minority groups and of the disabled.

A political programme based on the provision of need involves selection and priority. In the same way, support for certain innovations involves selection and priority. Politically- and socially-directed innovation processes are seen as a way of prioritising certain needs that can be identified through participation by users in the design and development of technologies.

'Needs' is a teleological concept in that it always assumes some end goal or purpose, that is, needs are inextricably linked to values. Obviously difficulties arise on how to define goals—is it a process of consensus or an explicit political programme? The GLC Economic Policy Group acknowledges that 'People have conflicting needs: sometimes because of different values and desires. An economic strategy based on social need would challenge the inequalities but it would seek to express and fulfil different values and desires' (GLC 1983, p. 29). This definition recognises the complexity of a social definition of need but the problem still remains of selection between needs and how to secure the articulation of need.

Further, production for social use/need is a difficult area to define practically—social use is context-dependent and it is perhaps easier to define what is not socially useful than what is. In terms of alternative technologies 'needs' are meaningful when there is some form of concrete possibility of realisation. For a programme of socially-directed innovation of alternative technologies problems exist in the sense that needs are related to the choice of solution on offer. For example, radiation treatment and chemotherapy are the choices on offer in the cure for cancer.

Questions that arise include: how it is possible to secure the articulation of need without a range of choices being already available; whether there are ways of offering alternative problem-solutions that can stimulate the articulation of needs; how, once alternative technologies are developed, it is possible to offer these to groups who do not have access to economic resources. Market 'needs' are assumed to be articulated as (economic) demand—is there a way of meeting needs despite the financial constraint on the public sector?

Conclusion

This chapter has sought to demonstrate that technology is socially constructed in that social factors—be they gender relations, class relations or economic relations, are embodied within technological artefacts. The identification of these social factors allows the possibility of the identification of alternative solutions to technological problems that can embody different social criteria—social use, environmental considerations, design for particular needs. Theoretically, it is possible to institute a policy for socially-directed technological innovation that recognises the processes of innovation and implementation as continually interactive for technological development.

In Part Two, the practical experiments of the 'radical' local authorities in Britain to formulate and implement a policy for socially-directed innovation are investigated. The discussion of the Lucas Plan highlighted some of the problems of this social approach to innovation—markets, access to finance, and so on. Importantly, the political commitment to change does not necessarily lead to different routes of technological development.

Noble (1983a; 1983b) argues that the missing ingredient in these approaches is the power to choose. Whilst alternative technologies are possible theoretically, the evidence of them practically is very thin. This is not an economic or technical problem but rather a political and social one: 'Without the requisite social power that would deem labour's alternatives viable—whether economic in present terms or not—they will be dismissed on economic and technical grounds—but for political reasons' (Noble 1983b, p. 81).

For a period in the UK, the 'radical' local authorities had at least an element of the necessary power.

PART II: Experiments in Alternatives

Introduction

The conclusion that it is possible to direct technological innovation and change towards particular political and social objectives is based on an understanding of technology as a social and cultural system. Societal choice is a necessary prerequisite of technological advance, although in many cases it may appear that the debate is about technical options rather than societal values. For example, arguments surrounding the choice of nuclear energy reflect opposing values concerning the risks to human life and the environment, rather than the technical superiority of the technology in question. In the same way, the identification of alternative solutions to technological problems that are based upon the criteria of human need reflects a political stance which seeks to challenge the notion of the desirability and feasability of the current path of technological development. In Britain the political commitment to the development of socially-useful products by the 'radical' local authorities represents a belief that the relationship between technology and society has become increasingly divergent.

The background to the policy initiatives on technological innovation formulated by the 'radical' local authorities, particularly the GLC, is informed by the view that technology is socially constructed. Product design reflects the dominant values of competitive market economies; social use, whilst a key component, is secondary to the expectation of economic return. In this way, by emphasising socially useful products, the 'radical' local authorities were presenting the possibility of an alternative path to technological development based on the primacy of social criteria rather than economic criteria. The social objectives are not disguised as technological options, instead the question becomes one of what kind of technology is required to assist in the fulfilment of these objectives. For example, Dr M. Cooley, Director of the Technology Division at the Greater London Enterprise Board (GLEB), argues that:

Existing uses of new technology all too often dispense with the skill, ingenuity and creativity of ordinary people. That is why GLEB gives special emphasis to developing new 'human-centred' technologies, which by contrast, build on human abilities, developing systems which respond to human intelligence rather than the other way round (GLEB 1985b).

In Chapter 3, a case study of the technological initiatives formulated and implemented by the GLC is described. It represents a practical experiment in the development of a socially-directed innovation policy. But before looking at this policy it is necessary to locate the 'radical' local authorities in the context of contemporary British politics. Although the GLC has been singled out for particular analysis, the other 'radical' local authorities mentioned, Sheffield City Council and the West Midlands County Council, have also attempted to develop alternative routes to technological development by emphasising the importance of socially-useful products. It needs to be borne in mind that the policy for technology forms only a part of overall industry and employment policies developed by these councils. Their overall strategies will be described in general terms before the focus is put on policies designed to stimulate technological innovation.

Experimentation with technological change is not just happening in the UK; in Chapter 4, other practical experiments in developing technologies in Europe and elsewhere will be described.

'Radical' Local Authorities

The experience of economic recession in the UK during the 1980s has been illustrated by rising numbers of unemployed, falling living standards and the decline of regional and inner-city areas. Economic decline affects and overextends the provision of services by the local authority, especially housing, health and the personal social services. As a consequence, local authorities have become increasingly concerned with efforts to develop their role in local economic regeneration.

The importance that has been attached to technological innovation as a key factor in economic development is evident in the growth of both central and local government involvement in schemes designed to promote innovation. For the main part these schemes, for example, Science Parks and Innovation Centres, look to the private sector to secure economic development. In contrast, the 'radical'[1] local authorities argue that this form of indirect intervention is not the way to meet needs or to solve the problem of unemployment and the decline of the local economy. What is required are forms of direct public intervention that look towards identifying needs, meeting these and creating jobs. Indeed, the key factor in the debate between the central Conservative government and the Labour-controlled 'radical' local authorities has been differing views on the role of the public sector and the way that the economy should be run. One form of this political battle has been played out in the context of the local authority: the legal right to intervene in the local economy.

In recent years the role of the local authority has emerged as a major political issue in the UK. This can be attributed to three factors. Firstly, since the late 1970s relations between central and local government have been characterised by attempts to control authority spending. Under the 1983 Conservative government this central issue is reflected in two major pieces of legislation: rate-capping[2] and the abolition of the higher-tier authorities in London and the metropolitan areas.[3] Secondly, the increasing importance of left-wing party policies within local government, that is, the renewed interest within the Labour Party in local government as an area of political struggle and socialist advance based on grassroots support. And thirdly, the increasing polarity between the policies of the major political parties, particularly in the area of public spending.

The 1981 local government elections saw the left-wing 'radical' councils, such as the Greater London Council (GLC), Sheffield City Council and the West Midlands County Council (WMCC), come to power in an atmosphere of excitement about the socialist potential of the local authority. The radical economic strategies formulated by these councils were seen as a vehicle for the mobilisation of support for alternatives to the economic policies of the central Conservative government. Michael Ward, the Chairperson of the GLC's Industry and Employment Committee, wrote: 'the Government is committed to the free play of market forces as the dominant principle of social organisation. Each successful Greater London Enterprise Board Project . . . is a demonstration that there can be [an] alternative' (Ward 1983, p. 16).

The 'radical' local authorities were concerned to demonstrate the importance of public sector institutions and organisations to the economy. The market-ruled society, and some form of democratic socialism, are the respective positions taken by the Conservative government and these Labour-controlled localities. These competing views on the future forms of economic and social change that are envisaged for Britain mean that the background to the economic initiatives of the 'radical' local authorities has been one of tension and conflict in the national political context. In the end, the GLC and metropolitan councils, including the WMCC, were abolished in March 1986. Thus, the only 'radical' local authority mentioned in this book which remains is Sheffield City Council.

Any discussion of the economic policies of local authorities needs to be informed by the fact that local authorities in Britain possess only limited economic and political power. Financial, legal (especially since April 1986) and geographical constraints, together with the myriad of assumptions, rules, regulations and procedures of local government all act to limit the possibilities for change.

The 'radical' local authorities referred to in this work are the GLC, Sheffield City Council and WMCC. A brief description of their policies will serve to illustrate their approach. 'Restructuring for labour' (Goodwin and Duncan 1986) is the term used to refer to economic initiatives which emphasise the importance of local community and workplace involvement in local economic development. The notion of alternative plans, socially-useful products, and production for social need are important components in the economic strategies developed by these councils.

The GLC

A Socialist Policy for the GLC was the title of the 1981 Labour Party Manifesto with which Labour took control of the GLC following the election. 'Jobs for a Change' was the slogan give to the approach which sought to:

1. bring wasted assets—human potential, land, finance, technological expertise and resources—into production for socially-useful ends.
2. extend social control of investment through social and co-operative ownership and increase trade union powers.
3. develop new techniques which increase productivity while keeping human judgement and skills in control (Editorial Collective 1982, p. 125).

Upon taking office, the new Labour group created an Industry and Employment Committee and an Economic Policy Group to assist the Committee in the formulation of the London Industrial Strategy (GLC 1985). The London Industrial Strategy (LIS) comprised a series of detailed sectoral reports on particular industrial, commercial and service sectors that built on the knowledge of union and user groups; tenants associations, community groups and so on. It was envisaged that the information would act as a base for the development of a strategy within which the GLC could intervene to help save and create jobs in London.

A second innovation was the establishment of the Greater London Enterprise Board (GLEB) as the main vehicle for the implementation of the strategy. Its major functions included: 'Investment to provide strategic or structural change to include municipal and public enterprise and industrial co-operatives; general investment in enterprises newly starting up, at risk of closure or operating in areas or trades with high unemployment; and the development of factory sites' (Boddy 1984, p. 171). The GLEB will be discussed more fully in Chapter 3.

The Co-operative Enterprise Board was also established with the specific task of encouraging and assisting new producer co-operatives. The Greater London Training Board was involved in advice on employment planning and labour-market problems and in the development of training schemes in relation to labour needs and the London Industrial Strategy.

The basic orientation of the approach was 'popular planning'; that is, the active involvement of community and work-place groups in the development and implementation of the London Industrial Strategy. This involvement took the form of work-place and local community plans that could begin to form the base from which 'wasted' resources (unemployed workers, empty factories and unused machinery) could be reintroduced into the local economy. Linked to this was the attempt to involve people in the technological innovation process and to develop socially useful products. A number of Technology Networks were established to facilitate this organisation of the innovation process by promoting links between the expertise in polytechnics and trade union and community groups developing alternative plans for employment. For example, the exploration of the possibilities for linking employment strategies to public sector heating needs was the basis of a 'Jobs from Warmth' plan.

Among the stream of initiatives adopted by the GLC to promote economic regeneration were: bottom-up popular economic planning; new forms of work organisation; planning agreements; co-operatives; socially-useful products and services and Technology Networks. This book focuses on initiatives related to technology and technological innovation. Whilst the case study in Chapter 3 focuses on the GLC and particularly the GLEB's initiatives in this area, reference is also made to Sheffield City Council and the WMCC strategies for technology.

Sheffield City Council

In May 1981, the Labour-controlled Sheffield City Council set up a new Employment Committee and Department. The objectives were summarised in 'an initial outline' document published by the Department in 1982 that stated the aims as:

to co-ordinate everything that the City Council can do to help:
1. prevent further job loss in the city,
2. to alleviate the worst effects of unemployment and to encourage effective training for new skills and jobs,
3. to stimulate new investment, to create new kinds of employment, and to diversify job opportunities in the city,
4. to explore new forms of industrial democracy and co-operative control over work (City of Sheffield Employment Department 1982, p. 2).

Like the GLC, Sheffield Council sought to develop a directly interventionist role in the local economy. The Employment Department, staffed by politically committed people from outside the traditional local government professions, began work in 1982 with a budget of £2.5 million. Eight project teams work on the following areas:

Research and Resource Team
To develop an early warning system for changes in the local economy, and to help prevent the loss of future jobs, through the development of alternative employment plans for key firms, industries and sectors, in dialogue with those who work in them and those who use their products or services.

Equal Opportunities (Women and Employment)
To investigate the employment situation and needs of women and to develop job-creating and (re)training proposals for positive action both inside and outside the local authority.

Training for Employment
To work with other agencies to develop a training policy and plan for key sectors of employment in the city; identifying specific future skill requirements within the local economy; promoting tailor-made (re)training programmes; and co-ordinating a critical response to the MSC temporary make-work schemes.

New Technology and Product Development
To monitor the impact of new technologies upon existing employment; to identify and generate opportunities for diversification into new, human-centred technologies, skills and jobs.

Aids to Enterprises
To offer financial assistance, specialist advice and premises to support existing and new firms, workers' co-operatives and job creation projects willing to negotiate a planning and economic agreement with the City Council.

Economic Development
To investigate the opportunities for directing large-scale investment in the local

economy, through co-financing arrangements with institutions (e.g. banks, pension funds, etc.) and the best organisational arrangements for managing such investments (e.g. local enterprise boards).

Municipal Enterprise
To explore opportunities for the local authority to generate jobs by an expansion of its own role as a local investor, trader and provider of services.

Industrial Development
To promote Sheffield industries and products nationally and internationally, by organising local participation in trade-fairs, together with marketing advice and assistance; and to attract new firms and new investment to Sheffield. (Reprinted in 'Struggles in the Welfare State' 1984).

Discussions with the women's and labour movements have led to modifications in policy; 'this is because the members of the department do not see themselves as laying down policy from above but as part of a process of development involving a wide range of groups and interests' ('Struggles' 1984). Since 1982, increasing emphasis has been given to public sector employment. In 'Strategies for the Employment Department 1983/4' (Employment Programme Committee, 1983), support for campaigns against expenditure cuts and privatisation, and other projects relating to local authority employment was given.

In relation to initiatives on technology, Sheffield City Council have a two-pronged approach. On the one hand, the Sheffield Centre for Product Development and Technological Resources (SCEPTRE) was established jointly by the Employment Department and the City Polytechnic in 1983. The aim of SCEPTRE is to support enterprises and co-operatives in the development of socially-useful products, processes and services which can satisfy social need and improve the quality of life. A central feature of any product development is the close relationship between producers and consumers. SCEPTRE offers specialist advice in engineering, metallurgy, marketing and business development and, in some cases, help with finance and premises. Since its formation it has:

Established a bank of product, service and process ideas;
Developed some product proposals to the prototype stage;
Assisted in prototype to full production activity;
Provided a general technical service for approved projects (*Sheffield in the Eighties* 1985, p. 16).

On the other hand, Sheffield City Council are keen to encourage the introduction of more conventional new technology. In the same report it is stated that:

> In co-operation with the Department of Trade and Industry, South Yorkshire County Council and the City Polytechnic, the City Council has helped to establish the Micro-Systems Centre in Sheffield to give advice on the introduction of new technology to private firms and co-operatives, and has supported other initiatives which ensure new technology is used to enhance employment and benefit industry.

As a further initiative, a land-development scheme adjacent to the university and polytechnic is under way for a Technology Campus. It is envisaged that this will provide a focus for industries using new technology. In a Presentation Summary to Sheffield City Council in September 1985, the objectives of the Technology Campus were described:

> The central objective of the Technology Campus is to provide the right environment, resources and location for the development of new technology industries. It is intended to encourage new industries and provide new employment, making research resources available and assisting an interchange of ideas and joint use of facilities.
>
> The aim is to foster a wide variety of research-based industries using new and existing technologies, particularly in electronics, computing, materials and biotechnology.
>
> It is envisaged that the Campus will include flexible workspace, a business development centre, product development workshops, high quality business units and accommodation for technology based industry (Sheffield Technology Campus 1985, p. 20).

Thus, whilst there is a commitment to innovation and the development of socially-useful products and alternative technologies, the adoption and diffusion of existing, more conventional, new technologies (for example, information technology) is also seen as an important component in economic development. The question of the possibility of changing the direction of technological advance through the development of socially-useful products is set against the realities of economic decline and the view that private sector uptake of new technology of the more conventional type is the way forward to prosperity. This dual approach is also evident in the policies of WMCC and is a question that will be addressed in the case study on the GLEB in Chapter 3.

West Midlands County Council

Similar to the GLC's creation of the GLEB, the West Midland County Council created the West Midland Enterprise Board (WMEB) in 1982. With a budget of £3.5 million, the main objective was investment in selected medium-sized companies which have potential for long-term growth and employment creation. The long-term nature of the WMEB's investment strategy is emphasised as central to economic development:

the positive commitment of the investor to the productive process rather than an interest in short-term money mangement. Through planning agreements, involvement of unions and organisations like the shop stewards' movements (with their ideas for retraining and building on workers' engineering and production skills), the investments would be part of a new approach to the planning of production with finance firmly linked to a long-term view of the productive process (Minns 1982, p. 96).

Like the GLEB, the WMEB was established as an independent company with Directors appointed by the Economic Development Committee of the County Council. It is financed under Section 137 of the Local Government Act 1972 which allows a 2p rate to be spent in the interests of the community; it is envisaged that pension fund finance, including the Council's own pension fund, would provide additional funding. The Board functions alongside an Economic Development Unit with responsibilities for industrial strategy (it is intended to base investment on sector analyses) and monitoring, as well as income maintenance, training, Council purchasing policy, co-operatives and other economic initiatives which relate to income support and redistribution. Although the main focus of the policy is on medium-sized firms, the support of co-operatives is seen as an important factor for the support of unemployed workers attempting to develop products and retain skills.

The WMEB approach to technology policy exhibits some commitment to the development of alternative technologies by its support for the Unit for the Development of Alternative Products (UDAP) located at Lanchester Polytechnic in Coventry. UDAP was established in 1981 as a result of the need for engineering and technical assistance on some of the product proposals in the Lucas Shop Stewards' Combine Alternative Plan (The Lucas Plan). The support for UDAP by the WMEB is based upon their commitment to the support of co-operatives; UDAP provides assistance to co-operatives and small enterprises wishing to develop alternative products.

However, in the main the technology policy of the WMEB is closely linked to its industrial policy. The main emphasis is on the utilisation of existing technologies rather than the design and development of new technologies. The WMEB's industrial policy focuses upon the regeneration of mature industries, therefore a major role for technology policy is to make both workers and enterprises more aware of the potential benefits of new technology. To this end a Technology Transfer Centre set up jointly with Aston University and an Advanced Manufacturing Demonstration Network, became operational in 1986. In an overview article on West Midlands approaches to technology transfer, Liff (1985, p. 5) defines technology transfer as 'a form of innovation which does not require the firm applying the new technology to undertake original research development work'. The main objective of the Technology Transfer Centre is to improve the dissemination of information relating to innovations that are relevant to West Midlands industries.

Although the technology policy of the WMEB is aimed at the adoption and diffusion of new technologies, there is an awareness of the potential adverse side-effects of deskilling and job loss. The Council Training Programme acknowledges the possibility of alternative utilisations of technologies that could benefit both workers and employers. The training programmes seek to break down sexual and racial skill divisions, provide training in new technologies and also develop two new areas:

—New Technology Awareness which gives workers and communities the skills to evaluate proposals for new technology and develop alternatives;
—New Technology Agreements. Advice to Trade Unionists in how to negotiate over issues concerning new technology. (Liff 1985, p. 8).

Conclusion

From this brief description of the economic and technology policies of the 'radical' local authorities it is possible to see how the priorities and style of strategies differed. For example, the GLC and Sheffield City Council adopted fairly radical approaches, taking on board many of the ideas about socially-useful production pioneered by the Lucas Aerospace workers' plan and emphasising socially-directed innovation via the Technology Networks in London and SCEPTRE in Sheffield.

While in some parts of its work the West Midland County Council adopted a similar approach via UDAP, there are important differences. Their policy has focused primarily on 'diffusion' (the provision of information on existing ideas) rather than on supporting invention and innovation. The main emphasis has been on technology transfer, for example, via the Technology Transfer Centre.

Even in West Midlands County Council's more limited involvement with innovation, it has tended to emphasise a more conventional approach, for example, the commercially orientated Warwick Science Park, set up jointly with Barclays Bank at the University of Warwick. In general, the West Midlands approach is less radical than that adopted by the GLC and the GLEB—with more emphasis on conventional corporatist definitions of what types of innovation should be pursued.

In the next chapter, the focus is on the activities of the GLC. This is not to suggest that the work of the other councils is not important or relevant, but the GLC's programme was the largest and most influential, perhaps inevitably since it involved the capital city.

Notes

1. The term 'Radical' is used here to describe the local authorities with a Labour left majority who were concerned to introduce public sector-led industry and employment initiatives.
2. Rates are local taxes which Parliament empowers local authorities to levy on the occupiers of land and buildings within their area. Rate-capping is a form of penalty introduced by the government; it refers to cuts in the Rate Support Grant provided by the government to local authorities that exceed their spending budgets.
3. The Local Government Act 1972 introduced a three-tier system of County Councils, Metropolitan County Councils and District Councils. The Metropolitan Counties were six councils that covered the densely populated areas of Greater Manchester, Merseyside, West Midlands, West Yorkshire, South Yorkshire and Tyne and Wear. Along with the Greater London Council, these metropolitan councils were abolished in March 1986.

3 The GLEB Innovation Experiment

The output of the innovation process is a range of new technologies, products, systems and processes—some are 'radical' in the sense of challenging existing technological trajectories and paradigms, often being based on scientific or technical breakthroughs, others are more incremental developments. Subsequently there is a process of deployment and diffusion—often accompanied by further incremental developments.

As has been observed, the 'radical' local authorities attempted to intervene at several points—the West Midlands, for example, focused more on the diffusion of existing 'new technologies' (for example, microelectronics, computers and robotics) many of which were also being diffused and deployed by conventional industrial and commercial interests.

The GLC and Sheffield City Council focused more on developing new products themselves—with a specific commitment to socially-useful products, although at the same time they were also keen to diffuse and develop any of the existing 'new technologies' which they felt were socially or strategically relevant. According to the GLC (1982, p. 2), socially-useful products are those that 'in manufacture and in use conserve energy and materials; [whose] manufacture, repair, and . . . recycling . . . can be carried out by non-alienating labour; and [that] should assist human beings rather than maiming them'. In this chapter an account is given of the experiment in socially-directed innovation by the GLC and the GLEB.

The GLEB was established as a limited company by the GLC in July 1982. It was financed, until the GLC's demise, primarily under Section 137 of the 1972 Local Government Act; this refers to the 2p rate to be used for the benefit of Londoners. Outside this, GLEB is free to obtain finance from other agencies in the public and private sector in hand with its investment programme for job creation and preservation. As an investment body, the GLEB acts as the main institution supporting the implementation of the GLC's employment strategy.

The operational guidelines for the GLEB were formulated by the Industry and Employment Committee of the GLC. The main objective was the implementation of the GLC's industrial strategy which sought to reverse industrial decline and stem job loss through the establishment of a capacity to intervene in and plan key aspects of the local economy. The role of the GLEB was to provide technical and financial support to enterprises and make relevant investment interventions that would further the development of London's industrial base, improve technological prospects and thereby preserve employment. An annual funding agreement reached with the GLEB by the GLC provided a budget of £30 million in 1982–3 and £25 million in 1983–4. The GLEB was thus, essentially, a publicly funded investment agency—somewhat similar to the British Technology Group (BTG). But it differed from BTG in that it was charged with acting on the specific policy recommendations established by the GLC; investment was to be based on both economic and social criteria with the creation of employment as the prime goal. Investment functions covered three main areas:

1. Investment to promote strategic or structural change. This refers to investments in industrial co-operatives, new public enterprise and municipal enterprise; areas which are central to the London Industrial Strategy (LIS) and which are unlikely to be supported by existing private sources of funds.
2. General Investment covers funding to both public and private sector enterprises within Greater London, however within this category priority is given to enterprises that are newly starting up, at risk of closure or operating in areas of high unemployment, whether this is a geographical area or a particular trade and where retraining opportunities are evident.
3. Development refers to site acquisition, factory building and other related activities. (Barratt-Brown 1984, Part 1).

The objectives of the GLEB are set out in their *Corporate Plan*—the linking of financial assistance to social priorities is a central feature of their strategy to improve the quality of jobs in London, to promote enterprise planning and industrial democracy and to ensure equal opportunities for women, ethnic minorities and the disabled:

Strategy
a) Sector Development
Where possible, GLEB will assist in the development of London's industries to benefit both consumers and workers. This will demand an understanding of the fundamental characteristics of each target sector, the drawing up of restructuring plans, and their implementation by specialist divisions through the GLEB's investment strategies.

b) Equal Opportunities
GLEB aims to tackle racial and sexual discrimination in all its projects. Enterprises will be encouraged to develop and promote equal opportunities policies and practice for women, ethnic minorities and people with disabilities. Conditions linking financial support to the achievement of these objectives will continue to be applied.
c) Novel methods of social ownership and control.
This will involve formulating for each investment or intervention an appropriate Enterprise Plan, the precise form of which will vary according to circumstances. In all cases trade union access will be required, and improvement in the quality of jobs will be emphasised.
d) New Forms of Property Development
Property initiatives will be pursued with the involvement of the local community, to provide infrastructural support for technology, projects and area initiatives.

Area Programme—GLEB has selected two areas as special initiatives. These are Hackney Road and the Royal Docks. Area Offices have been set up, staffed jointly with the GLC, to co-ordinate and integrate the full range of GLEB's programme within the two areas.
e) Acknowledging the acceleration of technological change, GLEB will continue to play a leading role in the promotion of socially useful applications of technology, and will actively encourage links between the academic and commercial worlds.
f) Company Restructuring
GLEB will play an exemplary role in restructuring and reviving companies in difficulties by the use of enterprise planning and the application and identification of entrepreneurial skills.
g) The Co-op Sector
Cooperatives will be fostered and supported because they represent one radical method of worker involvement in production.
h) Education and Information
GLEB will publicise its approach to the problems facing the London economy and the role Londoners can play, using methods which encourage two-way communication with the public (GLEB 1985/6, p. 6).

The Structure of the GLEB

To ensure the implementation of the strategy the GLEB has drawn on personnel from a wide variety of backgrounds. The board of part-time directors is drawn from the private, public and voluntary sectors, with knowledge and experience of business, finance, applied technology, local government, cooperatives and the trade union movement. The GLEB houses seven functional divisions:

Sector Strategy investigates particular industries of importance to the Greater London economy and develops strategies which guide GLEB intervention.

The Investment Division provides financial and business management appraisal.

Structural Investment concerns itself with enterprise planning, liaison with trades unions, and the development of co-operatives, new forms of ownership and control and industrial democracy.

The Area and the Property Division tackles the problem of dereliction and high unemployment in parts of London, and is developing an integrated system of property development advice in line with Enterprise Board policies.

The Technology Division sees new product proposals through from initial design to the establishment of manufacturing enterprises, and links academic research to groups in the community with ideas, through a series of technology networks.

A Finance and Administration Division handles the internal organisation of the GLEB, and an Information Division promotes all aspects of the Board's work through press liaison, publications and exhibitions (GLEB, 1983, p. 7).

Social Criteria

For the GLEB the primary emphasis is not a financial return on investment; projects and enterprises supported need to be viable, however investment appraisal includes the assessment of a social return to the community. The willingness of enterprises in receipt of GLEB support to enter into planning agreements, the provision of equal employment opportunities, the manufacture of socially useful products are examples of possible social benefits. This general approach is summarised by the GLEB's former director, Alan McGarvey:

We do require all our investment projects to be economically viable—otherwise jobs saved or created cannot be secure.

But we do not look for the maximum financial return.

Instead we look for a social return, in terms of the number of jobs—and equally important, their quality.

Our policy of enterprise planning seeks to guarantee work people basic rights—to trade union organisation, to basic information about their company, to equal opportunities, and to a voice in the future of their companies. Willingness to implement these policies is a condition of GLEB assistance, just as much as economic viability.

Projects which meet these two criteria are also judged by how they contribute to other social objectives.

Is the product socially useful? Is the project of special benefit to women or ethnic minorities? Is the enterprise a co-operative, or one that contributes in some other way to greater industrial democracy?

> If the answer to any of these is 'yes', the project has added weight.
>
> But we also follow strategic objectives for specific key sectors of London's economy, worked out in our sector strategy divison in co-operation with the GLC's Economic Policy Group.
>
> These strategies don't just provide a framework for judging the projects that apply to us for assistance.
>
> They provide a basis for us to go out and actively seek investment openings where we feel this would give us a vehicle for implementing our strategic plans for particular sectors. (GLC, 1984a).

The GLEB attempted to achieve these goals by investing in selected geographic areas, for example areas of high unemployment, and in selected industrial sectors. The approach may best be illustrated by looking at the type of intervention and investments that the GLEB has made.

Investments

GLEB targeted a number of existing industrial sectors sometimes uncharitably called the 'sunset industries' for investment designed to revitalize them—for example furniture, clothing and printing—by introducing more up-to-date equipment and improved products. But at the same time, as we shall see, it also invested heavily in the expanding 'sunrise industry' sector based on microelectronics and related technologies.

In parallel with these major technology-orientated investments, GLEB also invested in a range of workers' co-operatives, many of them serving 'social markets' or providing community services. The social criteria obviously stand out clearly in this context. For example, Lambeth toys is a co-operative producing educational toys that reflect the multicultural nature of London. Turnaround is a distribution co-operative specialising in books and pamphlets produced by independent, radical and community publishers that are often ignored by the major distributors; their range covers feminist, peace, gay, ethnic minority, industrial and environmental issues (GLEB, 1984b). Figure 3 gives a rough schematic representation of GLEB's projects based on its first full year up to March 1984. Although, as indicated, specific Divisions within GLEB were primarily responsible for specific initiatives (for example, Technology Division was responsible for the technology networks) there was also considerable overlap with the Sector Strategy Division providing strategic analysis and guidance.

Other major investments made by the GLEB were in enterprises facing difficulties or closure. The attempt to avoid job loss in larger

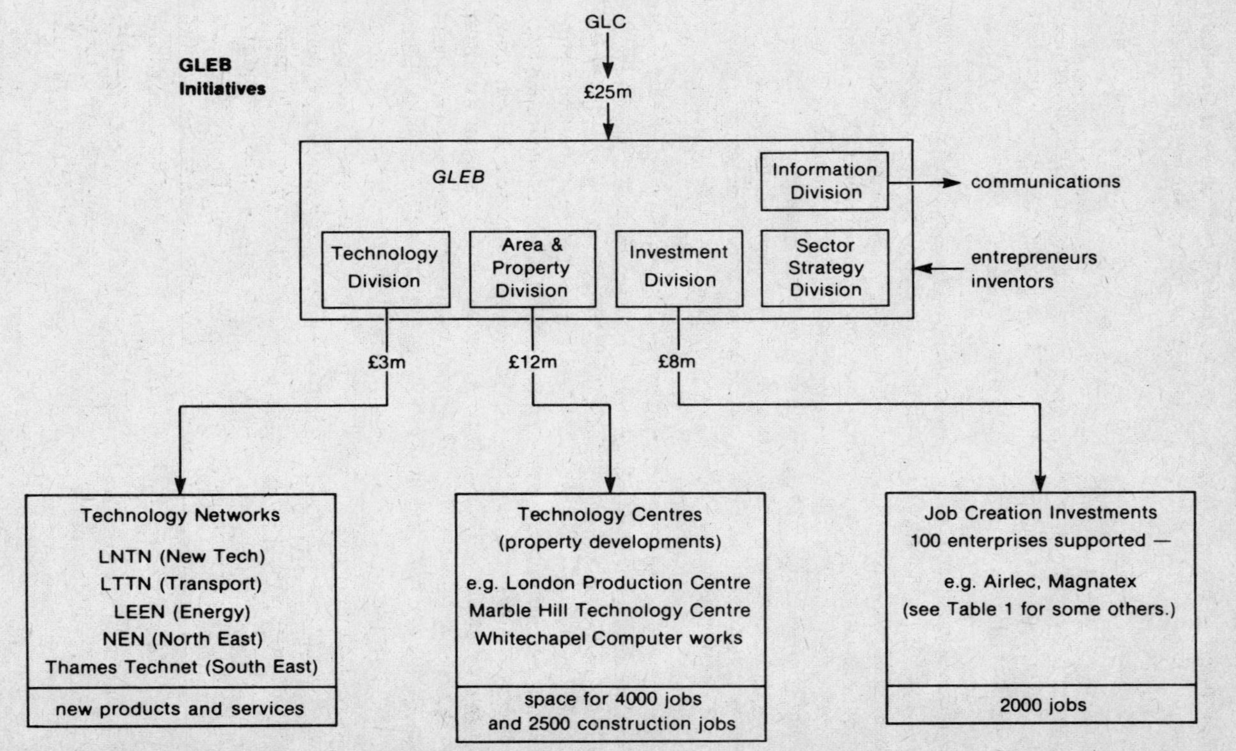

Figure 3 Schematic representation of GLEB's projects.

enterprises was a central concern for the GLEB and it was in this area that 'enterprise planning' and worker involvement in the company was emphasised. This is shown in the following case study (Barratt-Brown 1984, Part IV):

One of London's few remaining car component firms producing lamps and mirrors was reported to the GLC's Industrial Development Unit through its Early Warning System as being in trouble. The trade union convenor made a joint approach to GLEB with the manager who was equally concerned about the attitude of the owners. After lengthy negotiations both with the owners who were reluctant to sell and with the main customer, British Leyland, GLEB bought the company with a £90,000 stake in its future development, thus securing the employment of 200 workers. The company was not making losses but needed considerable financial resources to invest in the modern equipment and production systems essential for any firm to survive in this highly competitive market, which Japanese and other foreign producers have been taking over.

A new management board was established, made up of GLEB Directors and existing management, and a new enterprise plan is being negotiated between GLEB, the management and the Unions. The plan will include training, industrial relations, equal opportunity provisions, new products, new investments, future wage levels and price levels and a lasting collective bargaining procedure over these matters. This will be negotiated and discussed over a period of time, so that the workforce and its union representatives can be involved from the start in a new joint venture.

As a result, a strong trade union organisation has not only been saved but strengthened and made into the basis for the company's future development. With new ideas from the workforce and with technological back-up from GLEB's New Technology Network, it is intended to help the company to diversify away from its total dependence on the motor car industry. If the diversification is successful, it will put the company in a position to give a lead to others in this sector both in creating jobs and improving the conditions of labour.

In essence the GLEB is concerned to promote alternatives to unemployment. Worker involvement in enterprise planning and the stress that is put upon new products and manufacturing opportunities have much to do with the influence on the strategy of the Lucas Plan initiative and the development of workers' plans in other industries.

Product diversification as a means of regenerating enterprises is a factor in many of the GLEB investments. For example, Walter Howard Design, a furniture company in receipt of GLEB support, which was involved in high-volume production of low-priced bedroom and living-room furniture is to have its premises refitted and redesigned in preparation for a new phase of development. The new design-led product range that is planned is a bid to capture a share of the higher-quality modern furniture market (GLEB 1984b).

Another example of the GLEB support for innovation is Wagon Unloaders Ltd, a Chiswick company which is developing a novel form of freight unloading: the UndeRover. The product was devised in response to British Rail's call for a mobile unloader which could reduce the prohibitive costs of current freight handling which force many smaller companies to transport bulk by road. The GLEB acted to put the company in touch with AMD Engineering, another GLEB investment whose precision skills in engineering are important to the project. A prototype of the UndeRover, a single-person-operated, self-powered mobile unloader with could handle solid fuel, grain, minerals, chemical powders and other dry goods, was developed with the help of British Rail Research and the National Materials Handling Centre at Cranfield. The £50,000 investment by the GLEB was matched with a £50,000 grant from the Department of Industry under the Government's Support for Innovation Scheme (GLEB 1984b).

To facilitate innovation and the promotion of socially-useful applications of 'new' technologies, the GLEB have adopted a two-pronged approach to technology initiatives. The first are essentially property developments, technology centres or parks housing small new technology enterprises where it is hoped that the clustering of similarly orientated companies in the electronics sector will provide a more secure path to the longer-term development of these companies, for example the London Production Centre in Wandsworth and the Richmond 'Synergy Park'. The other initiative is the promotion of technology transfer from academic institutions to local industries, and the encouragement of worker and community involvement in the innovation process, via a series of 'Technology Networks' established by the Technology Division of the GLEB. It is this area, the GLEB's activities in both new technologies and socially-useful products, that forms the bulk of this case study. It must be recognised that this involves highlighting and focusing upon one particular area within the GLEB's overall strategy for industry and employment.

Technology Initiatives

Mindful of its limited financial resources, the GLEB approach to technology is strategic in the sense that its response to technological opportunities is to provide 'seed-corn' finance to projects at an early stage of development. It is at this point in their development that difficulties are encountered when attempting to secure funding from traditional sources—banks, venture capital agencies, and so on.

The development of technologies and start-up technology enterprises at an embryonic stage is seen to require a 'nurturing' strategy. The Technology Networks are designed to give back-up support and facilitate technology transfer from academic institutions to local enterprises. The Technology Parks, it was hoped, would then provide a secure environment for start-up enterprises.

Within the Technology Division the selection of technological projects to support is guided by the general economic and social criteria for investment of the GLEB. Particular emphasis is given to socially-useful products (for example, energy-conserving projects and products for the disabled) and exemplary projects which extend technological opportunities to otherwise disadvantaged groups, for example training in new technology for women and ethnic minorities.

The Technology Networks and Technology Centres and Parks are seen as exemplary projects by the GLEB which indicate alternative responses to the stimulation of technological development. Priority is given to projects which have links with the local community and the local borough, which involve disadvantaged groups, which are co-operative ventures and which have technological implications.

Together with the commitment to novel socially-useful products, recognition is also given to the need to introduce more conventional microprocessor technology and its applications and other developing technologies into various industries in London. The technological areas that are delineated as appropriate for the GLEB and Technology Division support are areas which do not have high entry costs, are not capital intensive and do not use technologies which degrade or destroy skills and knowledge.

In essence, the GLEB's role in technology initiatives is to stimulate the appropriate development of technology with 'seed-corn' finance at an early stage of project development and to facilitate the development and promotion of socially-useful products and technology transfer through the Technology Networks. The next section looks at the mechanisms that were established for the achievement of these objectives.

Technology Networks

Innovation theory points to the importance of the coupling of technology-push (technical possibility) with market-pull (demand) as the basis for the successful development of new products. In most cases, innovation initiatives are only indirectly concerned with job creation—the rationale being that commercial success and growth will inevitably lead to local employment. The GLC and GLEB approach explicitly identifies technological innovation in the form of socially-useful products as a means to create local employment.

The provision of back-up support for the development of enterprise plans by existing firms facing collapse, newly established enterprises and co-operatives and to aid product diversification in existing companies was given by the GLEB in the form of a series of Technology Networks linked to local polytechnics. The basic aim of the networks was to tap the technical resources of polytechnics so that technical advice could be given to client enterprises. They have been established as public companies with their own board of directors drawn from associated organisations. The idea has some similarity with the privately managed 'Science Park' concept, except that rather than the pursuance of purely commercial goals, the GLEB is attempting, via the Networks, to transfer technological expertise and ideas from polytechnics to the community with the creation of employment as the prime goal. The GLC and the GLEB thus saw the Networks specifically as a preferred alternative to Science Parks.

The Background to the Technology Network Idea

In the Labour Manifesto leading up to the 1981 local government elections that saw the election of the 'radical' Labour-controlled GLC, special emphasis was put on 'production for need', referring specifically to the Lucas Aerospace workers' campaign for the right to work on socially-useful products. It was stated:

> Groups of workers such as the Lucas Aerospace Shop Stewards' Committee have, with the support of the Labour Party, began to develop ideas on alternative production—using technologies which interact with human skills; making goods which are conducive to human health and welfare; working in ways which conserve, rather than waste, resources.
>
> We believe that these initiatives—which constitute a fundamental rejection of the values inherent in capitalist production—must be supported by a Labour GLC. We shall therefore be prepared to assist groups of workers seeking to develop alternative forms of production, with finance, with premises, or in other ways (Labour Manifesto, Greater London elections, 1981).

The background to the technology network idea is directly related to the

Lucas Combine's establishment of the Centre for Alternative Industrial and Technological Systems (CAITS). The Centre grew out of the need for facilities for work on the product prototypes put forward in the Lucas Alternative Plan, the rationale being that the more prototypes of alternative products designed and developed to meet the social needs of both the producer and the user, the stronger the bargaining position of workers fighting redundancies.

For CAITS the use of the resources within an academic institution, North London Polytechnic, allowed the possibility of bringing together different design and technical skills which often were not found together in one work-force. This co-ordination of skills made possible the production of alternative products on a local or regional basis rather than only within companies employing a high proportion of design and technical staff (Wainwright and Elliott 1983).

The combination of academic theoretical skill with 'tacit' skill, 'the things we know but cannot tell' (Polanyi, 1976, p. 79), that is, skill which is built on experience, is a key factor within the CAITS and Technology Network idea. The concern was to create the facilities within which both types of skill could be acknowledged, exercised and developed.

The commitment to socially-useful products and alternative technologies provides a means by which to, firstly, relate technology to social need through products such as aids for the disabled etc., and secondly, to present a vision of an alternative paradigm that prefigures a different role for technology in society. Of major importance is the ability to influence, in some way, the direction of technological advance and the benefits that flow from it. To do this it is necessary to produce both a critique of the current shape and aims of existing technologies together with examples of alternatives that could lead to social and technological change. The base of this alternative paradigm that is being fostered by the GLEB's Technology Network initiative is the use of resources; materials, energy, capital and labour, to meet social needs in a more effective way than they are currently catered for. The ability to transcend existing forms of thinking about technology by building upon tacit knowledge and experience, what M. Cooley; the Director of the Technology Division at GLEB terms the 're-integration of the hand and brain', is seen as the way to stimulate more socially relevant and technically sensible ideas. The Technology Networks were established to provide the facilities for this alternative approach to technology and innovation.

The Industry and Employment Committee of the GLC in its 1982–3 Capital Programme for GLEB included a provision for expenditure of up to £3.2 million on technology. It was proposed that the Council's concern to develop new technology would be most effectively implemented by the establishment through the GLEB of a number of area-based and product-based Technology Networks incorporating innovation centres, product banks, machinery/production/equipment banks and technical support and advice units (GLC 1982).

In their General Policy Document (GLC 1982) the GLC acknowledged the need to look at the problem of technical change in many of the industrial sectors in London and to attempt to harness the appropriate use of new technology to the creation of employment. The support for new technologies was based on the view that 'human-centred' technology, building on human skill rather than displacing it, was the preferred direction of technological advance. It was stated (GLC 1982, p. 1) that 'new technology can be designed to maintain and develop existing manual skills and tacit knowledge in a way that enhances work rather than degrades it'. Alternative technologies should be in the form of socially-useful products, as defined on p. 55.

Thus, while the GLC and GLEB recognised the importance of technology to the development of the London economy, the emphasis was put on 'alternative ways in which [it] can be introduced, which have quite different implications for energy and material use, for ecology, for those who operate the new technology and for those who use or consume its product' (GLC 1982, p. 2).

Two central propositions informed the GLC approach to technology: firstly, that 'one of London's most precious assets is the skill, the ingenuity, the creativity and the sheer enthusiasm of its people'; and secondly, 'London possesses in its three Universities and seven Polytechnics one of the richest scientific resources in the world' (GLC 1982, p. 2).

Against this background of thought on technological alternatives, and influenced by the success of CAITS, the creation of five Technology Networks was proposed. The networks would exhibit the following features:

a) they would be based in separate buildings or in a distinct area of an existing Polytechnic or University College, and be staffed by people who act as the interface between the professional scientists and the community. They would be sited and designed in order to welcome those who would normally be hesitant to enter academic institutions, perhaps having a shop front as the Science shops have so successfully done in the Netherlands.

b) they would include a new enterprise support unit which could provide technical, production, planning and marketing advice to co-operative production and other similar initiatives, drawing on the expertise within the scientific institution, as well as the skill of their own small staff.

c) the small interface unit would arrange for appropriate contact for people in the community to make use of the facilities of the polytechnics and universities for their own research and development needs.
d) the centres would be intentionally sited in localities rather than being concentrated at one place in London in order to widen the access of people to them. At the same time they would be able to draw on GLEB's central technological resources for specialist help.
e) the networks would also encourage undergraduate and post-graduate projects on product ranges which would be useful to the communities around them.
f) we would expect some specialisms to develop between the networks, determined in part by the facilities and special expertise of the institutions to which they were linked (energy on the South Bank for example), and in part on the needs of the area (the need for mechanical and engineering technology in the West London area).
g) each network would contribute to a product bank which would consist of a pool of product innovations which would be patented for use by working people. An Innovation Bank of a similar sort has been developed in Limerick in Ireland, sourced both by product ideas produced in Ireland as well as those imported from abroad. The British Technology Group (BTG) also have available product proposals which may be licensed by enterprises. In the case of the technology networks, we would expect ideas to be contributed from the educational institutions, from technical and manual workers in the workplace and communities, as well as being brought in from abroad.
h) networks would also include as appropriate a machine bank, consisting of second hand machinery which had been refurbished as part of a training programme and was then available for use by co-operatives and other appropriate groups.
i) the networks should aim to involve in their management and day to day operations not only representatives of the scientific community, and of locally based workplace groups, but also representatives of the local community and other voluntary groups concerned with the development of socially useful technology (GLC 1982, p. 3).

The IEC Report on Technology Networks has been described in full because it provides a good account of the basic idea of a technology network—in theory. The creation of organisations to promote the appropriate use of technology to satisfy needs that are unmet by the market offers an example of an approach to technology transfer, product development and design which is democratically controllable and open to the influence of trade union and community groups. In essence, the technology networks were formed to promote socially-useful technologies. A central feature is participation by groups interested in developing alternative products; in this way there is a recognition of the importance of the social process of innovation. This point was made in the GLC's newsletter,

'Jobs for a Change' (GLC 1984a): 'GLEB's Technology Networks enable people for the first time to turn ideas about alternatives into prototypes of actual products.'

The idea of the Technology Networks rested on two main propositions: user involvement in product development; and links between innovation and employment. This meant that in practice the networks differed quite substantially from the CAITS model which had clearly defined and focused activities in both a political and practical sense; it provided information and technical assistance to workers developing alternative plans in particular industrial sectors. The Networks were more loosely defined and faced the task of building their own political constituencies.

Five Technology Networks were established, two were area-based and the other three were technology- or issue-based. The area-based networks are Thames Technet, the South-East network which is linked to Thames Polytechnic and the Product and Employment Development Network for North and East London (PEDNEL). This latter network is now known as the London Innovation Network; it has links with the Polytechnic of North London and the 'Design Development Unit' (DDU) at Middlesex Polytechnic. The three technology- or issue-based networks are the London Energy and Employment Network (LEEN—focusing on energy related equipment and services), the London New Technology Network (LNTN—focusing on microelectronics and information technology), and Transnet (focusing on transport issues and related technologies).

The five networks have both trade union and community group representatives together with representatives from voluntary sector groups and the polytechnics on their management committees. They operate as independent organisations with their own buildings, administrative staff and workshop facilities. The funding of projects is via Technet Ltd, a company formed by the Technology Division at the GLEB. The Management Committee, known as the Network Council is responsible for policy decision-making within the general guidelines formulated by the GLEB, administrative staff are responsible for day-to-day management.

It was envisaged that the Technology Networks would provide the research and development facilities, using the polytechnic resources if necessary, for the development of prototype products which could feed into the GLEB's wider strategy of support for new and existing enterprises. The product prototypes would then be produced and manufactured by GLEB-supported enterprises and co-operatives.

The technology networks represented elements in an essentially supply-side strategy (technology-push) to be met by the 'market' of unmet needs (demand) articulated by the community and work-place groups involved in a process of local planning and enterprise planning. The 'market' was seen to comprise of public sector service provision together with the possibility of the commercial exploitation of products.

The major concern for the networks in their early stages was the formulation of an organisation and focus that could begin to address the problem of how to promote socially-useful technologies in a competitive market economy.

The wide brief of the Technology Networks meant that their practical implementation depended upon perceptional interpretations by those directly involved in the establishment of the networks of the focal point necessary to achieve their objectives the promotion of socially-useful technologies, user involvement in product development and employment. As each of the Networks are described the differences between them and the strategies adopted to acheive these objectives are made apparent.

The London Energy and Employment Network (LEEN)

LEEN was the first network to be established; it was formed in 1983. The brief of energy and employment was very broad and posed difficulties for the initial activities of the definition and establishment of a coherent organisation. The idea of a network is a federation of groups that come together to work on a particular project; this definition raised the problem of organisational structure. LEEN began as an umbrella organisation for a number of groups interested in energy-related issues, these included London NATTA, ITDG, SERA, ERR, and representatives from the South Bank Polytechnic's Centre for Energy Studies. In addition, the Central London Polytechnic was involved via its support work for the establishment of an Energy Conservation and Solar Centre (ECSC) which gradually took a leading role in LEEN. Local Heating Action Groups, the Warmth Energy Saving Co-operative and the Lewisham Energy Centre became involved at a later stage. At the outset LEEN had an open form of management and functioned as a funding body for different organisations involved in particular energy projects. The main focal structures were the London Energy Centre at Avonmouth Street which housed workshop facilities for product development and the ESCS at St Pancras.

Following this period it was decided by the GLEB that a more corporate structure was required with full-time employees if LEEN were to develop a co-ordinated policy and strategy. In this way the network shifted away from a loosely co-ordinated open structure to a solid organisation with a strong central core. From July 1984 the reorganisation meant that LEEN employed its own staff and had a work programme that was in part to do with job creation and energy projects and partly organised as a consultancy.

The membership of LEEN is comprised of organisations who elect a management committee. Full-time staff are largely composed of people with technical qualifications and experience.

This discussion of the organisational changes within LEEN highlights the problem of developing a working management structure for the networks that is based upon a clear definition of strategy. It was found that the rationale for the establishment of the networks, the promotion of alternative products and the provision of access to workshop and technical facilities leading to socially-useful employment was not the main problem regarding energy related issues discovered by LEEN. In the field of energy at least at the local level the main factor is not the lack of socially-useful technologies; rather the technology exists, but what is required is the political, institutional and financial commitment to the redistribution of resources that would allow the implementation of these technologies.

In consequence, the main focus for LEEN has been to work with local authorities to campaign for central government investment in energy efficiency and conservation measures while at the same time taking on consultancy work and providing advice and information for councils and tenants. For example, the London boroughs of Hackney and Islington have introduced 'Right to Warmth' campaigns. LEEN provides back-up support in the shape of heating advice services to tenants, housing associations and the local authority. These heating and insulation 'packages' consist of a combination of different conservation measures appropriate to particular housing stock, for example, tower blocks in Hackney and old people's homes in Southwark. To support this work LEEN has also developed computer programmes and a new type of low-cost data logger for use in energy audits in public buildings. The exemplary use of these measures in Council housing provides a model of what is required for London's housing stock. The use of this model in the borough's campaigns for extra finance from central government to allow the implementation of conservation measures in Council housing adds potency to this process. That said, LEEN did press ahead with a wide range of new product ideas.

Product development in the LEEN workshop facilities was geared towards commercially viable technologies that were also deemed to be socially useful depending upon their use and the particular market areas they were aimed at. For example, Bill Wright, an inventor from Reading University, contacted Leen and ITDG with an idea for a small and economic 100W wind turbine. The product could be marketed as a domestic electricity source, and in addition has an important use for developing countries, and particular interest was attracted from Vietnam. Once the development work at the LEEN workshops was completed an independent company, Remote Power Applications Ltd, was established.

Another example is the 'Pedelec Stella', an electric bicycle invented by Phil Campopiono and developed in the LEEN workshop with financial assistance from the GLEB. The bicycle was formally launched in March 1985, just after the launch of the Sinclair C5 electric tricycle, and a company, Stellmar Ltd, was established for production. Phil Campopiono, who received £76,000 for development work, stated: 'Without GLEB we would never have done it. Nobody else would have touched us' (*South London Press* 25 January 1985).

The GLEB had been keen to support the electric bicycle concept, not simply because it seemed likely to be commercially successful and would provide employment but also because it could make cycling more viable for elderly and infirm people, people carrying shopping loads, and so on. The GLC were attempting to extend the 'cycle-way' programme in London and generally to support the use of bicycles. Electric power-assisted bicycles were felt to be more environmentally appropriate than petrol-driven 'mopeds', and were cheaper to buy and run.

Other projects supported by the GLEB and the LEEN included a novel steam engine for a motor launch developed via ITDG, and a domestic heat pump system developed by LEEN. Two prototypes of the domestic heat pump were developed and tested in 1984 and ten units were scheduled for production in 1985 for field testing. LEEN also developed a 'user-friendly heating controller', and a ventilation control unit. The LEEN workshop facilities were used for further development on the road/rail bus prototype which was initially built at CAITS using a conventional bus that was modified to be able to run on both road and rail. This was one of the projects that had emerged from the Lucas Aerospace workers *Alternative Corporate Plan*.

It was hoped that the products and projects that were developed in the workshop at LEEN with investment finance from the GLEB would, with a working prototype, be more likely to attract finance for production and manufacturing from more traditional private sector sources such as banks and venture capital agencies. In this way the GLEB and

LEEN's role was of nurturing embryonic projects with socially-useful and employment potential by supplying the initial investment research and development facilities.

But as we have seen LEEN was not only concerned with developing new products. Increasingly it found that the emphasis of its work shifted to the deployment of existing, fairly simple, energy-saving measures—such as draughtproofing, backed up by advice, information and monitoring services. Indeed LEEN found that the *main* strategic need was not to develop new technology but rather to campaign for the resources to enable existing systems to be deployed.

The question of how to promote appropriate and socially useful technologies was addressed in different ways by the networks, depending upon their technological and geographical areas.

London New Technology Network (LNTN)

LNTN was established with a focus on microelectronics and information technology. Its aims are to encourage the application and development of 'new' technologies in a socially useful way. 'New' technology refers to those areas of recent rapid development including microelectronics, computer technology and software development, and its spinoffs in the areas of information technology, robotics, automated manufacturing systems, and so on.

An important aspect of the LNTN's function is to 'demystify' the new technologies by making the facilities and personnel available to trade union and community groups and running educational programmes in the skills required for using the new technologies. In addition, the Network is concerned to promote equal opportunities in employment in new technology areas, for example, it houses a women's new technology co-operative (Live Wires) and runs an EEC-funded microelectronics training course for women.

The exploration of the social uses of computers and computer-related technologies is centred around the belief in the need to democratise information and pool information resources for use by community groups and other organisations. The LNTN provides general advice and services to the other networks concerned to apply new technology to their products and services. For LNTN the interest is in the applications of new technologies rather than in invention and innovation of the hardware. Innovation is focused on computer software and the development of socially-useful computer programs.

The LNTN is housed in a 20,000-square-foot building in Camden. Up to half of the space is for small factory units for start-up enterprises involved in new technologies, for example, 'live wires', 'rolec music systems', and an enterprise involved in 'interactive literature' computer software. In addition there are workshop facilities for technical projects funded by the GLEB and an exhibition/meeting area which is important for LNTN's approach—the accessibility of new technology to workers and the community.

Projects The main area of 'exemplary' projects is in the field of 'expert systems'. It is envisaged that the computerisation of information on, for example, diabetic care and other medical areas could be of use to general practitioners in community health care—this project is in conjunction with St Thomas's Hospital and the City University. Another area for the socially-useful application of new technology is in the link-up between small inexpensive computers used by community groups to facilitate the pooling of information. A project involving the development of cheap computer networks for domestic and community schemes links St Mary's College computing expertise with LNTN.

LNTN acts as a back-up to enterprises, some of which are in receipt of GLEB assistance, for example at the London Production Centre, one of GLEB's technology centres. LNTN is involved in work on control engineering for educational robots for a company set up as a spinoff from Imperial College. The work is helping to develop LNTN's control engineering expertise, and is relevant to further development of educational robotics products, emphasising the 'human-centred' approach to systems design.

Human-Centred Technology A major area of the GLEB's interest is in the growing field of human-centred technology. Dr M. Cooley, Director of the Technology Division at GLEB, had been a pioneer in this field in the United Kingdom together with Professor H. Rosenbrock of UMIST (University of Manchester Institute of Science and Technology), and it figured strongly in the 1976 Lucas Combine Committee's *Alternative Corporate Plan*.

The case has been made by Cooley (1983) and others that new machine and data-control systems using computers could be designed to enhance workers' skill and creativity rather than displace it and even lead to redundancies. In the context of Computer-Aided Design (CAD) Rosenbrock (quoted in Cooley 1983, p. 28) argues:

> We need not develop CAD systems which refuse to use the special skills of the operator and the special properties of the human mind. We can instead develop systems which will accept the skill of the user and collaborate with it to increase

productivity. We need not develop flexible manufacturing systems which fragment and destroy the machinist's skill. We can allow that skill to develop into something new. The same can be said of office automation and, to the extent that it is not too late, to printing. Such professional areas as medical diagnosis will offer the same kind of alternative choice. This I regard as the most important challenge facing engineers and technologists in the next 20 years.

Rosenbrock continued to pursue research in this field at UMIST, and GLEB, with the involvement of LNTN, maintained a strong interest. In 1986 they were awarded £1.8 million by the EEC to oversee a human-centred manufacturing project under the EEC's ESPRIT programme in liason with UMIST and other research institutions.

For LNTN the promotion of socially-useful applications of new technologies has centred around attempts to demystify these technologies by providing training and allowing access to interested groups. The understanding of the possible alternative applications in computing, and so on, which can enhance the quality of work is seen as important for support to trade union groups in bargaining for New Technology Agreements and alternative processes of implementation of new technologies in the work context. An important aspect of their work at LNTN is the development of prototype alternative applications.

Transnet

The third technology- or issue-based network is Transnet, which is focused on transport technologies. Whilst an important role for the Networks is in the support of trade union organisations fighting redundancies, the development of alternative products which can offer constructive alternatives to unemployment takes time to build up. In contrast, Transnet developed more directly out of a campaign against the closure of the London Transport Bus Repair Works at Aldenham and Chiswick.

This was the latest network to be established, and was formed in July 1984. The aim of Transnet is to encourage socially-useful and responsible planning and implementation of transport policy and the development of socially-useful technologies to fit in with this policy, for example, the Pedelec Stella developed in the LEEN workshops.

The consideration of the socially-useful potential of products is based on their overall design in the context of a socially-conscious transport policy. The emphasis is upon addressing the problems of transport service provision for the disabled, low-income groups and women without incurring large costs. The network rejects the promotion of high-technology transport in preference for simplicity and effectiveness. The capital-intensive nature of support for high technology is beyond the Network's financial capability and, more importantly, does not fulfil the social objective of employment retention and creation.

The founding of the Network in response to the resistance to redundancies at the London Transport works meant that Transnet had a firm focus on alternative production to bus repair and maintenance. As part of the campaign to restore economic viability to the Aldenham and Chiswick works, the partly completed chassis and tools for the Ward Chassis, a multi purpose vehicle chassis, was bought from the receivers of Ward Motors Ltd. This new product was aimed at small bus and coach manufacturers because of its low floor, easy maintenance and low cost, and its manufacture would serve to retain jobs at London Transport Engineering.

The support for alternative plans by workers forms a key aspect of the networks and the GLEB's role, and the potential effectiveness of this form of bargaining for alternatives to unemployment is exemplified in the above example. However, once London Transport was no longer the responsibility of the GLC but of central government in the form of London Regional Transport, talks broke down, largely as a consequence of the introduction of government legislation for LRT which prohibited third-party manufacture in their engineering works.

Transnet has been organised as a network which devotes three-quarters of its resources to campaigning and the promotion of socially-useful transport technologies. Product development accounts for the other quarter.

Projects The network's support for the development of low-level as opposed to high technology is evident in the involvement in work on a steel hull work boat that lasts longer and requires less maintenance than the fibre-glass models which are currently favoured by boat builders. The market which still exists for steel hull boat is filled by the Netherlands, but Transnet argue that this project could bring work and employment to the London area. Support for this project is linked to the campaign for a freight policy that looks toward rail and water rather than road transport.

A small firm using the LEEN workshops is working with the Polytechnic of the South Bank on 'Hush-kits' modifications to meet the requirements of the GLC lorry ban. A further project is the development of an Underground information presentation for blind people.

Transnet is concerned that its promotion of socially-useful transport technologies and their campaign for the better use of existing transport facilities, for example the canal system, form an integrated approach to transport policy. The network recognises the importance of looking at transport policy from the point of view of disadvantaged groups and environmental concerns; for example, pollution and traffic congestion.

Thames Technet

Thames Technet, which has links with Thames Polytechnic, is an area-based network with its own innovation workshop which is accessible to a wide range of potential user groups. The concentration on building close links with community groups is a prime consideration of this network.

As an area or geographically based network, Thames Technet covers the Greenwich, Lewisham, Southwark, Lambeth and Bexley areas of South London. The aim is to act as a practical resource for individuals, groups, co-operatives and other enterprises on the development of products and services that are relevant to the needs of the local community. The workshop facilities are designed to house a number of start-up units for local enterprises.

The Network was formed in 1983 and has links particularly with the Mechanical Engineering and Continuing Education Departments of Thames Polytechnic. The present building houses, in addition to the workshops, a canteen and a creche. The organisational structure of Thames Technet emphasises a strong community bias both on the Management Council and amongst the full-time staff, who are divided into Technical and Community Teams. The Technical Team have specialist skills and are required for the operation of equipment in the workshops. The Community Team are responsible for liaison work with the local boroughs, the assessment of the resources and needs of the area and the building of links with community and voluntary organisations.

The main aim of the Network is to link technology with the community in the sense of making the technological resources of the network accessible. The focus on accessibility is seen as the means by which local people can begin to reassess views about the nature of technology and its relevance and use to their daily lives. For example, is technology relevant to the needs of local community groups and would a computer-based information package be a valuable resource for these groups?

Whilst the Thames Technet is committed to the questioning of the efficacy and desirability of current technological trends, the difficulty is recognised of finding the staff with the correct mixture of skills, that is, with both technical expertise and a social and economic interest in technology issues. The reassessment of the 'technical' requires a reassessment of gender divisions (the 'technical' is often synonymous with male domination) and the divisions of labour. The problem at Thames Technet is how to overcome these divisions which are reflected in their own Community and Technical Teams.

Projects The workshop facilities are used to provide informal training in the use of simple equipment, for example welding machinery. The facilities are free to small start-up enterprises, especially co-operatives, for the first three months. Examples of the types of enterprise housed are a small business involved in the adaption and design of fashion clothes for the disabled and outsize women; and Blackwell Products, a co-operative making a low-cost aerobic composter and other gardening and recycling products, which has now moved into its own premises but still uses some of the resources at the Network, for example, computerised mail order. Two catering businesses have been set up in the canteen at the network, one of which is now applying to Lewisham Council for help with premises and funding. Thames Technet is concerned to act as a resource and catalyst for local enterprises which are initially housed and then helped with feasibility and development work leading to a business plan with which to approach the local council Economic Development Unit or Co-operative Development Agency. Thus Thames Technet is concerned with employment creation and helping enterprises to get started.

The links with community organisations serve to yield product and service ideas. In response to the needs of the disabled representatives of Greenwich Action for the Disabled there are plans for creative play equipment for handicapped children and a quick, efficient wheelchair repair and modification service.

The South London Children's Scrap project, aimed at identifying the scrap availability in the area, is a community project that could be converted into a product or service.

The Toys Workshop Venture is a community-based enterprise, brought together during its development stages, via Thames Technet, Greenwich Employment Development Unit, Greenwich, Lewisham and Bexley Toy Libraries and Spurn House Resettlement Unit. The aim of the enterprise is to provide a range of appropriately designed and produced toys which take into account the needs of able-bodied and disabled children.

In terms of community health, Thames Technet is working with Dulwich Hospital on the development of a mobile vascular unit that could act to alert individuals of the possibility, in the early stages, of arterial diseases.

A project with Thames Polytechnic is on mobile bulk-handling equipment for unloading ships which may have important Third World applications.

For Thames Technet the focus is on socially-useful products that are directly community-based and have employment potential. The close relationship with the community allows the identification of needs that are not necessarily met by the development of new technologies. The important linkage is the appropriate use of technological resources for community organisations and the encouragement of a critical attitude to technology issues.

The London Innovation Network

LIN acts as an umbrella organisation for the Product and Employment Development Agency for North and East London (PEDNEL), the Design Development Unit (DDU) at Middlesex Polytechnic and the Community Construction Design (CCD) with links with North-East London Polytechnic. PEDNEL seeks to encourage local community and trade union participation in LIN, whilst the DDU, with its commitment to socially-useful products via student projects, and the the CCD are design- and product-orientated. Thus LIN represents an attempt to combine area-based and technology-based organisations.

LIN houses workshop facilities for prototype development and small-scale batch production. The DDU was set up at Middlesex Polytechnic in May 1984 using the resources of the Craft, Design and Technology Department and the skills of the students to provide a design and prototype service to the Network. Design is identified as a key point in the innovation process for intervention to direct product development towards social needs; of particular importance is design flexibility so that products can be adapted and modified to meet particular needs.

The work of LIN is directed towards the creation of employment through 'technology transfer', the linkage of product ideas from academic institutions or other groups in the community to local enterprises. This involves the identification of product needs and the support of projects at the prototype and product-development stages. LIN acts to assist the final stage of technology transfer by seeking new enterprises or co-operatives to manufacture the products.

The emphasis on design and the use of design to accommodate particular needs represents a commitment to socially-useful products but also to the social process of design. LIN is keen to support projects that detract from the view that inventors are largely white and male (see, for example, the electronic aid for the deaf described in GLEB 1985a). The importance of educational provisions for craft, design and technology teaching is seen as a possible direction for the Network in its concern positively to involve the local community in the design process.

Employment creation via new product development is an area where the commitment to socially-useful production meets the necessity of commercial and economic viability if enterprises are to survive. There is care to ensure that products meet a special need but are also generally attractive to secure a wider market. For example, Playbox is designed to alleviate the problems of inadequate storage space in creches and nurseries. The wooden box unfolds and unpacks to become an assortment of different play shapes, for example a seesaw. Work began on the box in November 1984 and it is hoped that this product, along with other wood-based designs will be sufficient to establish a small woodworking co-operative in the area.

Projects Include an electronic aid for the deaf and a low-energy house, in addition, LIN has a major commitment to the project 'Disability and the Domestic Environment'. The project began as an investigative outreach programme in the Autumn of 1984. The contacts made with groups revealed the need for mobile repair services for aids for the elderly and disabled. The project established by LIN provides an example of the type of local authority service that is required if there were the commitment of resources to this area. As well as services there is also an emphasis on product design for the disabled where a lack of provision exists for particular aids and housing modifications. The aim of the project is to involve interested groups and build a picture of the needs of the disabled that would provide valuable background information for product development. The identification of gaps in existing aids and equipment availability, together with research into the problems of the disabled community, can then be used to initiate projects within the Network that can serve to raise the awareness of the needs of the disabled and begin to explore channels through which they can be met.

The emphasis on technology transfer from the DDU and CCD, as well as projects developed in the Network, to local enterprises presents certain problems with the promotion of socially-useful technologies, which are shared in varying degrees by the other Networks. Firstly, there is a need to develop marketing and entrepreneurial skills if enterprises are to survive within a competitive market. The recruitment of staff with a mixture of these skills together with a commitment to socially-useful technologies has proved to be difficult.

Secondly, the availability of manufacturing opportunities, without incurring large costs in terms of equipment and premises, is a difficult area. There are only 24 manufacturing co-operatives in London, mostly in labour-intensive sectors. The development of links with the commercial manufacturing world means addressing the problem of whether the networks are subsidising the research and development work of the private sector. Employment creation within the London area has suffered when the problem of manufacturing opportunities has meant that it has been necessary to approach companies established elsewhere. For example, the Whitfield Bench, a technology bench aimed at the educational furniture market that was developed by the DDU and LIN, could only be manufactured by a company outside London.

Thirdly, the access of resources from the polytechnics has proved to be very dependent upon the personal commitment of individuals. And finally, community involvement in the social process of design and innovation is not an automatic process once the facilities have been created.

Technology Networks—Summary

The technology networks were established to provide access to the facilities necessary for the enouragement of a socially-directed innovation process that was based on the close links between producers and users. The development of socially-useful technologies was seen as an important aspect of the GLEB's employment creation programme.

The strategy adopted by the GLEB of the provision of 'seed-corn' finance in the early stages of product development was seen as the key point of intervention in the design and innovation process. It was at this stage that design flexibility could be inbuilt to products to secure their accommodation of social needs. The facilities for research and development of product prototypes, the transfer of product ideas from academic institutions to local enterprises, and the provision of back-up support to start-up enterprises were areas around which the networks were orientated.

The objectives of the promotion of socially-useful technologies, user involvement in product development and employment creation were addressed by the networks in different ways. For LEEN it was found that the prime consideration for energy issues was the provision of information and advice on energy efficiency and conservation measures plus technical work wherever possible. Employment creation could best be obtained by campaigning to change policy on energy issues that would lead to the political and financial commitment to implement the socially-useful technologies that already exist.

For LNTN, the exploration of the socially-useful applications of new technologies, particularly in the area of computing, and the development of prototype alternative applications was the main focus. Open access to the network's resources, equal opportunities in training programmes were important to the aim of demystifying new technologies. Employment creation was in the form of 'nurturing' small start-up enterprises.

For Transnet, the integration of a socially-conscious transport policy with the socially-useful technologies meant that campaigning work was the largest area of activity. As in the case of LEEN, employment creation could best be achieved by changing attitudes.

The above networks were centred around products and issues. The area networks developed differently because of their close links with the local community; the emphasis was particularly on the identification of appropriate technologies for use by local groups and local enterprises. Thames Technet was perhaps the mostly closely involved with community-based projects, and was concerned to promote socially-useful technologies by encouraging a critical attitude to technological development. The 'nurturing' of small start-up enterprises housed in the network was seen as important to the creation of employment.

LIN exhibited the characteristics of both a product-based and an area-based network. The central focus for this network was on design and the transfer of design products from academic institutions to local enterprises. The problems that were encountered in facilitating this process will be discussed more fully in Chapter 5.

The technology networks acted as catalysts for the transfer of knowledge and ideas into employment creating projects; the particular emphasis was on the prototype and product development stage of the innovation process. The next stage of moving into production has proved to be the area where the GLEB involvement in the form of premises and finance is crucial as the following case study illustrates.

An innovation cast study: Airlec

In 1983, Chris Ross, a designer, had an idea for a fully-automated machine to load freight and luggage containers into wide-bodied airplanes like jumbo jets. This new loader would save energy because it ran on batteries, would cut down pollution, which is a serious problem at airports, and was the first machine of its kind to be run entirely by a computer. He did the basic drawings for his scheme, worked out costings, developed it as far as he could, and then looked for the cash to build a working prototype. Ross takes up the story of what happened then:

To develop the idea I needed to set up a company for which I needed large capital resources. Having approached many financial institutions I discovered that they were either unable or unwilling to properly investigate the idea as such to identify whether it had commercial prospects or it was even technically feasible. However we went to GLEB and they were prepared to investigate it at some depth and decided that it was sufficiently viable to invest considerable sums of money to take it through to producing a prototype machine.

Once GLEB had provided their support it was strange that the other financial institutions who we had previously approached and who had declined to investigate the project then became keen to investigate and in fact one of them commissioned Cranfield to look at the proposals who gave us a glowing report (GLEB 1985a).

GLEB provided £50,000 equity investment to enable the project to get started in early 1984 and Airlec Vehicles was established as a limited company.

When it came to finding premises, GLEB's help was also crucial. As a new company, Airlec found that landlords wanted 100 per cent guarantees for premises, which the working directors were not able to provide. GLEB's Property Division solved the problem by taking a lease itself, and subletting to Airlec.

GLEB also made it possible for Airlec to obtain a key executive by sharing his services on a consultancy basis with another GLEB-assisted company.

The first twelve months were spent making and refining the prototype, the last six in setting up the marketing operation and building the first production models. Prototypes were tested successfully at Heathrow and Gatwick and discussions were held with major British, American and Canadian airlines, the New York Ports Authority and a number of British provincial airports. Ross comments:

If it was not for GLEB's faith in what I was doing this scheme would never have got off the ground.

Because GLEB was prepared to back me—not just with words but with cash—I could go to others and they would listen seriously and then come up with what I needed. Before GLEB invested £50,000 in me no one else was prepared to do anything. I would go to industrialists or banks and they would look askance at the idea. They thought it was too big a gamble because it has to be built and made and is not really a fancy 'sunrise' product (GLEB 1985a).

In the Autumn of 1984 Airlec Vehicles' freight-loading machine was on show for the first time at a special exhibition at Gatwick Airport. Airlines throughout the world showed strong interest in it. Subsequently some firm orders started to come in, and early in 1985 the first Airlec 1070 fully-automated computer-controlled aircraft loader went into service at Liverpool International Airport. By mid-1985, the company seemed well on the way towards breaking into the £50 million-a-year market that Ross claimed was open to a product like Airlec's.

On the basis of the sales prospects, a refinancing programme was sought with major city financial institutions to prepare for full-scale production, on a basis which values GLEB's initial £50,000 investment at £500,000. A total of 100 jobs are expected to be created (GLEB 1985a).

However, this next stage, finding finance to move to full-scale production, is the hardest, especially during a period of financial constraint. Evidently support was not forthcoming and the Airlec project was stalled, a similar fate befell the Pedelec bicycle. We will be returning to look at the implications of failures like this in Chapter 5. GLEB's response has been to look to the 'technology centres and parks' in the hope that production of such new products could be started up there. It is to the work of these centres we now turn.

Technology Parks and Centres

The commitment to alternative plans and socially-useful products forms one-half of the two-sided approach to technology adopted by the GLEB. The Area and Property Division were also concerned with the development of a technological base via property developments. The emphasis of this side of the approach was on new technology and the so-called 'sunrise industries' based on microelectronics, computers and robotics. A series of Technology Centres were established, usually based in and around an existing high-technology firm. For example, a Synergy Centre was established in Richmond, based on AMD Engineering, a firm involved in robotics; the London Production Centre was set up in Wandsworth, based on a Rediffusion electronics plant, and the Whitechapel Computer Works was established in the East End. These were essentially property developments in that the GLEB bought the premises, provided support to the existing enterprise and offered 'nursery' unit space for other small enterprises. The expectation was that the Technology Networks, particularly the London New Technology Network, would provide back-up technical support for the new enterprises.

The Technology Centres represent an attempt to intervene in the innovation process from the technology adoption and diffusion side, as well as new product development. This diffusion aspect was a key component of the West Midlands Enterprise Board strategy, particularly in the area of manufacturing technologies, and was also evident in the Sheffield City Council's decision to establish a Technology Campus.

The general guidelines for investment by the GLEB still operate in the Technology Centres where each new enterprise is subject to the GLEB's technical and social criteria, with production for need and work-force involvement in planning as an essential part of the investment package. In this way the Technology Centres differ from the more conventional 'Innovation Centres' established by many local authorities which run on conventional market criteria. For example, the Centre established in North Kensington, based on an old bus garage, was specifically earmarked for Afro-Caribbean enterprises, with an emphasis on producing aids for the disabled.

The Synergy Centre

The Richmond 'Synergy Centre', known formally as the Marble Hill Technology Centre, was established by the GLEB in April 1984. The Centre is based around AMD Engineering, a company rescued from receivership by a GLEB investment of £205,000. AMD is a firm of precision and product engineers specialising in high-quality components and comprises the core company on the 28,000-square-foot site. It is hoped that the site will house a number of similar electronics and high-precision engineering companies to become a manufacturing centre, particularly in the growing 'sunrise' industry of robotics.

Operating from the same site, and a major customer of AMD, will be a company founded by John Reekie, described in *New Scientist* as 'the Henry Ford of robotics'. He has pioneered the production of cheap robotics from standardised parts, including robotic teaching aids, and a computer-controlled robot arm. His firm is now working on the production of a small tracked vehicle for bomb disposal, fire-fighting and similar uses (GLEB 1984d).

The 'Synergy Park', says Dr M. Cooley, the Director of GLEB's Technology Division, 'will ensure a working community of interdependent high technology companies, sharing facilities, striking sparks off each other, and ensuring that their work, and the subcontracting it will produce, all stays in London'. In particular the high-technology firms would be 'able to use the skills of AMD's craft workers for their specialised prototype and production engineering requirements' (GLEB 1984a).

The tenant enterprises will share wages administration, a canteen, a creche, swichboard, and human-centred Computer-Aided Design facilities. Tenants will also be required to implement GLEB's policy of Enterprise Planning. For example, the 'Enterprise Plan' negotiated with AMD Engineering and the relevant trade union (AUEW), comments that 'the company will always seek to produce products which are socially useful. No products will be manufactured which will be used in aggressive acts of war, or are capable of being so used.' In addition there are provisions for worker participation and access to information via consultative procedures and the establishment of an elected union 'worker director' on the company board.

Whitechapel Computer Works

The concern to support the new computer hardware manufacturing industry in London led to investment in Whitechapel Computer Works Ltd, a start-up enterprise involved in the manufacture and prototype development of the MGI, a 32-bit workstation with a very high-resolution graphics capacity. The American computer journal *Byte* called it 'the first truly personal workstation'; it has particular relevance to Computer-Aided Design applications. The company is based in the GLC-owned Whitechapel Technology centre.

The London Production Centre (LPC)

The London Production Centre is the outcome of a GLEB investment of £4.5 million in a two-acre site in Wandsworth. Rediffusion Radio Systems Ltd's move to Crawley meant the loss of 100 jobs for those workers unable to make the move. The GLEB scheme has meant that these 100 skilled workers have now been placed in a number of small companies operating on the site. The specialised companies include Broomhill Electronics, which makes circuit boards, wiring harnesses and other equipment, and companies involved in activities ranging from plating and spraying, packaging and shipping to printing and maintenance. The remaining space of around 70 per cent of the site will be converted by the London Production Centre into workshops and office and studio space, with a special emphasis on uses related to the electronics industry (GLEB 1984c).

The London Production Centre represents an attempt to support strategic industries like electrical, electronic and instrument engineering in London. The commitment to a 'nurturing' of small companies gives a priority to premises. Andy Hartwill of GLEB states: 'We've already got places to put very small companies when they're just starting up but, at some stage, you have to go into volume production—thats where LPC comes in' (*Electrical Review* 7 December 1984). The technology networks provide facilities for small batch production in start-up enterprises; it was envisaged that LPC and the companies operating there could provide further manufacturing opportunities for several satellite design and marketing companies with prototype products. This strategy would solve the problem of links with the commercial manufacturing world that were voiced by the staff of LIN.

This technology transfer strategy is planned for certain products developed through the networks, for example the microelectronic-based energy saving controllers for domestic and industrial heating and ventilation developed through LEEN. In the London Technology Strategy the eventual objective is seen to comprise four to six product design and marketing companies based on the London Production Centre site clustered around Broomhill Electronics as a manufacturer.

The adoption and diffusion of 'sunrise' technologies supported by the GLEB emphasise the manufacturing opportunities for products in the areas of computing, electronics and robotics. The Technology Centres act to nurture companies in their embryonic stage by providing premises, facilities and the possibility of on-site subcontracting work. It is envisaged that this strategy will lead to the establishment of high-technology manufacturing centres in key industries in London.

Conclusion

For the GLC and the GLEB the experiments in alternative production or socially-directed innovation have centred on the linkage between GLEB investments based upon social criteria, product development in the technology networks and the 'nurturing' of small enterprises that exhibit novel forms of social ownership, for example co-operatives and enterprise planning. In this way the GLEB has attempted to address the problems of social needs by directly linking needs to technological opportunities. Whilst this has meant the support for 'low-level' technology, there is also a commitment to the exploration of the social applications of new technologies and high-technology product development in the form of electronics and robotics.

The promotion of socially-useful technology is based on the view that technologies are socially constructed. Product development guided by certain social criteria—social use, environmental considerations and design for particular needs—allows the possibility of the identification of alternative solutions to technological problems, for example, in the form of human-centred technologies.

For a socially directed innovation process, the emphasis is upon the close relationship between producers and users, and user involvement in product development. It was envisaged that the creation of the technology networks would provide the facilities for the technology transfer from academic institutions to the community, but equally importantly provide a conducive environment for the exercise of the 'tacit' knowledge of work-place groups.

The innovation process involves idea and product development through to adoption and use. For the GLEB experiments the product development stage is an area of relatively low costs; the next steps of manufacture and production prove to be a difficult area for socially-useful technologies. These problems are discussed in Chapter 5. The question is raised as to whether wider social and economic factors mitigate against the possibility of socially directed innovation.

In the next chapter, other experiments in technological alternatives in Europe and elsewhere are discussed.

4 Other Initiatives

The case study of the GLEB innovation experiment illustrates an approach to the development of socially-useful technology that links innovation with employment, bases investment on social criteria of need, employment quality and equal opportunities, and encourages user involvement in product development. In this way the attempt is made, through exemplary projects within the Technology Networks and in the application of new technologies, to present the possibility of an alternative path of technological advance that is firmly embedded in the social milieu and which gives priority, over the economic benefits, to the social benefits to be derived from innovation.

The main objective of the GLEB approach is to forge a direct link between technology and social need, for example in the form of employment or heating needs. User involvement in the design and development of socially-useful products locates the source of invention and innovation within the context of implementation and use, the point being to demonstrate the possibility of choice and alternative solutions to technological problems. The discussion of the Technology Networks in the case study highlights the importance that is given to the transfer of technology from academic institutions to the local community, but in addition recognises the importance of other forms of knowledge, 'tacit' or experiential knowledge, to the innovation process.

The problems encountered in the Networks relating to the marketing and production of socially-useful products will be discussed later. However the technological initiatives for alternative development that are described in this chapter have been selected, firstly, to point to the variety of initiatives, but also, secondly, to show how similar problems have been encountered and addressed in Europe and elsewhere.

The GLC and the GLEB have been concerned to publicise their policies and encourage similar initiatives overseas—the promotion of international links has been seen as part of the strategy. John Palmer (1986, p. vii), director of the GLEB's Information Division, has pointed out that a dozen public sector bodies in as many countries are now involved in studies or projects to set up GLEB type bodies.

GLEB type policy and initiatives are easier to make in some countries than in Britain. Apart from the greater development of the worker co-operative movement, the legal system abroad means that communal and regional bodies already possess some of the developmental power which British councils seek.

In Australia several Labour-controlled administrations at city and state level have shown interest in the the GLC and GLEB approach, and it has been particularly emulated in Melbourne. In New Zealand, an Enterprise Board is being established in Wellington.

Several European countries have shown an interest in the establishment of bodies similar to the GLEB; France (notably the Val de Marne Department) and Spain (Barcelona) and West Germany (where the decentralised powers of the existing *Länder* give them obvious advantages). Greece has already established a Pan-Hellenic Local Development Agency, heavily influenced by the GLEB model.

The Maltese government has been particularly keen to discuss the GLEB's approach to job creation, with requests for assistance in the establishment of worker co-operatives. In addition, the technology networks, particularly in the area of the development of natural energy resources, have been approached for technical advice.

A large number of other organisations have also expressed interest in and visited the GLEB, it was reported in *Enterprising London* (GLEB 1984b, p. 3) that:

Among others, there have been visits from Credit Unions in Canada interested in applying personal savings to job creation and social investment. From Germany delegates from the Regional TUC for the Ruhr area visited to seek advice on the creation of a GLEB style investment body for their area, and the newly elected German Social Democratic Party administration for the Saar region came over with a similar request.

Latest German interest has come from the Green Party, which is looking to GLEB for advice on the construction of its own economic development policy. And this autumn a major delegation of local authority leaders from Barcelona in Spain is to visit the Board and some of its projects. Meanwhile interest from Australia, China, Denmark, New Zealand and Sweden, excited by earlier contact with GLEB, continues to bring casual and official visits from journalists, trades unionists, economists, and local and national government officials.

In addition to acting as hosts to visiting delegations and researchers, the GLEB has attempted to develop international contacts and links on a formal basis. The 'Technology Exchange' programme, established in 1985, is designed to ease technology transfer both within Britain and overseas.

Funding has been obtained from the EEC Commission for a joint programme with the Regional Development Authority for the Flemish Brabant Region of Belgium, which includes the capital; Brussels, and the Regional Development Authority for the Basque region of Spain; the area which includes the famous Mondragón co-operatives. This collaborative technology transfer arrangement will seek to find new product and process technology for small and medium-sized businesses in each of the three regions. The Technology Exchange is designed to match particular product innovations to the production facilities and market experience of specific companies. In addition, it acts to identify and encourage partnership arrangements for firms interested in generating extra income by licensing out their technology to overseas companies operating in markets that are not directly accessible.

The Intermediate Technology Development Group (ITDG), together with the GLEB, established the Technology Exchange as a means to give access to companies to inventions and innovations on a 'non-exclusive licence' basis, thus the fee is in proportion to the success of the product or process. The Exchange is in the process of compiling a computerised index of the available sources of licensable technology on the market, which will include the 'product bank' of products originating in the five Technology Networks. The director of the Technology Exchange, Brian Padgett, commented on the way the Exchange works:

> We work with firms on a client basis to meet their needs, building up a client profile based on their current technology and range of products and markets that they face. Then we explore with them the adjacent areas they could move into. Armed with that, we go through our sources to give them a filtered list of suitable opportunities (GLEB 1985a).

For companies in receipt of GLEB assistance, the Technology Exchange service is free; other firms may use it for a fee.

The second major international initiative funded by the GLC was the Third World Information Network (TWIN), established in 1985, and its offshoot, TWIN Trading. The Information Network and trade links are designed to promote mutually beneficial trade and technology exchange between First and Third World countries. For example, links have been established with Vietnam, Nicaragua and Mozambique. The aim of TWIN Trading is to foster links between London and the Third World based upon reciprocal exchange agreements; there are products and machinery that are made in London which are of use to Third World countries, likewise there are products made overseas which could find a market in London. For example, a delegation from Vietnam was keen to obtain tea-making machines in London (possibly from a GLEB-assisted enterprise); interest was also expressed in recycled machine tools and second-hand heavy goods vehicles. The idea was to arrange some form of barter; machines from London in exchange for local products from Vietnam such as rubber and coffee.

As another example, links were established with Cape Verde. In a Report to the Chief Economic Adviser of the GLC, Olivier Le Brun explains the project:

> Machine tools: The aim of the project is to give the opportunity to skilled workers who are using old machine tools, mainly lathes, to come to London to work for a few weeks in GLEB Networks on good second hand machines that they will adapt to the needs of their plant and go back to Cape Verde with the machines and a better understanding of the technology. The EEC has agreed to finance the travel expenses of the workers and the training in London. The machines could be bought by some kind of barter: tuna fish or perhaps clothes from the big, very modern Morabenza factory. Malta has offered a switch deal: to pay us the cost of the machines in exchange for tuna fish from Cape Verde (Barratt Brown 1985; Part IV).

This brief description of the international links that have been established by the GLC and the GLEB serves to demonstrate the possibilities for technology exchanges and transfer that are geared to building on the needs of London and the countries in Europe and the Third World for their mutual benefit. The interest in the GLEB model of an Enterprise Board rests largely upon the strategies developed to link innovation and employment.

Having briefly reviewed the diffusion of GLC/GLEB ideas in Europe and elsewhere, we now look at some case studies to explore the diffusion process and parallel developments in more detail—focusing first on the UK and then moving overseas.

SCEPTRE

SCEPTRE is an organisation, similar to the Technology Networks in London, that was established by Sheffield City Council Employment Department to act as a resource centre and interface between Sheffield

City Polytechnic and the local community. The focus of the centre is on product development and the provision of technical advice and support to local enterprises, especially co-operatives, and community-based organisations.

For example, the Advanced Humidifier System is a product which grew out of a joint venture between the Polytechnic, tenants' organisations and a local co-operative established by some of the workforce after the closure of a large firm involved in the manufacture of machine tools. The local authority housing department, under pressure from tenants to act on problems of condensation in its housing stock has provided an initial market.

SCEPTRE acts to support local enterprises throughout the process of start-up by providing technical support for product development, the provision of premises within their sheltered workshops and marketing and organisational development support. Particular attention has been paid to marketing where the links with the Polytechnic's Economics and Business Studies departments have proved valuable.

The access to the resources within the Polytechnic, both expertise and facilities, has involved a period of negotiation and are very dependent upon individual commitment and the political orientation of the particular department. Where expertise is required, SCEPTRE aims to tap specialist activity that is complementary to the expertise within the unit. The product ideas that have emerged within Polytechnic departments have tended to be rather crude prototypes. SCEPTRE acts to create something more tangible that is geared toward the market and particular needs by applied engineering before the product can be manufactured.

UDAP

For UDAP, the links with Lanchester Polytechnic Engineering Department are more developed than the SCEPTRE links with Sheffield City Polytechnic. The West Midlands County Council supports UDAP in a bid to foster the manufacture of alternative products in the West Midlands. UDAP acts to process project ideas from the local community; where specialist engineering expertise is required, student projects are initiated within the Engineering Department. UDAP is concerned to provide marketing and business planning support for local enterprises and co-operatives; to this end links have also been established with the Business Studies and Graphics Department of the Polytechnic, for student projects working with client enterprises.

UDAP is also involved in work with some Third World countries on particular projects; for example, they are using their workshop facilities to develop a brick-making machine for Zimbabwe and are working with the Eritrean Relief Organisation on problems of energy sources. In a similar way to TWIN Trading, local enterprises are linked with Third World countries requiring certain products. For example, delegates from India have expressed the need for a workshop built in a container which could be fitted in Britain and then shipped abroad. UDAP is seeking EEC funding for the project, which is successful would mean that a local engineering firm would design the workshop, buy and refurbish second-hand machinery that is reliable, serviceable and with easy access to spare parts—a problem of existing machine tools in the Third World.

Science Shops

The role of science in the community has been a subject of controversy throughout the 1970s, fuelled by environmental organisations and other groups who have argued that science has become elitist and removed from the social context. The Technology Networks and similar organisations in Britain have attempted to bridge this gap by drawing on applied expertise in the form of product ideas. In Europe, the 'Science Shops' have been geared more towards access to, and provision of scientific and technical information to, the public and community organisations that are seeking answers to particular problems. Their goals are, 'to provide a means for members of the public to seek answers to scientific and technical questions arising from their daily lives, and for scientists and engineers to apply their knowledge, training and skills to topics of social concern' (*Science* 1984).

In the Netherlands, the University of Amsterdam established the first official Science Shop in 1977. The University provides the premises, funds the employment of fifteen staff members and contributes to expenses. Since opening the Science Shop has dealt with numerous requests for information particularly in the areas of environmental problems, health, housing and factory working conditions.

The French 'Boutiques de Science' were instituted by François Mitterand and his socialist government; there are now seven in France including one in Paris and one in Marseilles. The idea grew out of a series of planning meetings that brought together trade unions and environmental groups previously excluded from discussions on science policy. The aim of the 'Boutiques de Science' is to integrate science and technology into society and fuel the debate about the role of modern science in society. John Stewart, the British biologist who is involved in the Boutique de Science at the Jussieu campus of the University of Paris, and is chairman of the Fédération Nationale des Boutiques de Science et Assimilés, states: 'Science shops are unique tools for making scientific culture something more than just the icing on the cake: it is a way of helping ordinary people use science as part of their daily concerns' (*Science* 1984).

The problems that have been encountered in experiments designed to link academic institutions with the community have, to a certain extent, been shared by both the initiatives in Britian and Europe. Firstly, the attempt to break down the barriers between scientists, technologists and lay people has proved to be quite difficult, particularly in areas of the need for applied research work where answers to problems do not fit in neatly to disciplinary areas but rather require an interdisciplinary approach. Secondly, there is the danger of a 'technological fix' attitude to social problems. And thirdly, there is a danger that the 'Science Shops' and other initiatives will act to reinforce community deference to scientific expertise. The technology networks, with their emphasis on the encouragement of the development of 'tacit knowledge', go some way to overcoming this last problem.

The Science Shops initiatives in Europe concentrate on the dissemination of scientific and technical information for individuals and community and workplace groups. The Technology Networks and similar initiatives in Britain are more concerned to tap academic resources for product development that can lead to employment creation. So although the aims of integrating science into the community are similar, the Science Shops do not encounter the problems relating to the production and marketing of alternative products that are present within the 'technology network' approach.

Marketing and Distribution: Mehringhof

One of the problems voiced by the London networks, but which is also a concern of other initiatives, is the marketing of alternative products. Socially-useful products are geared to the fulfilment of a need, but at the same time this need must take the form of a market if the product is to be manufactured. For alternative organisations this means competing in terms of capitalist economics; for example, the Lucas Plan identified the need for a domestic iron lung but the costs of producing it proved too high for the product to be commercially viable. The concentration on start-up enterprises by the Technology Networks means that they have been very product-orientated; marketing is initially a non-productive activity. Whilst alternative production in Britain has taken the form of product development, alternative production in West Berlin has taken the form of marketing and retailing.

The growth of the 'alternative movement' in West Germany during the 1970s, which incorporates a diverse range of left-wing political strands, has as a common thread the emphasis on 'alternative experiments in the present in order to demonstrate future possibilities. Mehringhof comprises two five-storey factory and workshop blocks which, though collectively run, house a number of independent projects primarily in the area of 'alternative' marketing and distribution of goods and services. The only project which is involved in production is an engineering collective working on research and development projects for wind-power stations. The concentration of projects in the service sector makes Mehringhof fairly typical of the majority of alternative projects in West Germany.

Production for social need, in the context of Mehringhof, is seen as the production of social relations, cultural values and attitudes rather than purely confined to actual commodities. Thus there is a bias amongst the projects toward 'cultural commodities' in the form of books, journals and alternative arts and leisure. The profits are given over to other alternative projects or to the alternative community to which these projects belong. Erica Carter (1985, p. 190), in an article on the Mehringhof, gives an example of what this means in practice:

> [one example is] a campaign conducted by Okotopia, a retailing and publishing co-operative in Mehringhof, whose range of goods includes Nicaraguan coffee, wine from various European production collectives, and tea imported from India and elsewhere. In 1982, Okotopia began a tea campaign which aimed to raise the share of third world producers in profits from retailing in first world countries.

The alternative marketing initiatives of co-operatives like Okotopia depend upon the support of politically conscious minority groups who are prepared to travel to buy products and produce, in this case, tea, coffee, wine and bread, which support their political beliefs and sympathies. As Carter (1985) states: 'more than price, it is political sympathies which determine consumer choice on this alternative market'.

In terms of mainstream marketing vocabulary, socially-useful production in the context of Mehringhof and other initiatives in West Germany, fulfils the needs of politically conscious groups. The emphasis on the production of 'social use values', as opposed to economic reward means that alternative projects suffer (sometimes severe) financial constraint. In a bid to alleviate some of these problems a number of alternative finance networks have been established throughout West Germany.

The Netzwerk Selbsthilfe (Self-Help Network) is based at Mehringhof in West Berlin and has been in operation since 1978. It offers alternative financial and organisational assistance to collective projects and political initiatives in West Berlin. It is financed on the basis of donations and monthly subscriptions from individuals and groups, and funds are then distributed in the form of loans or grants to projects which exhibit particular criteria; for example, the Network dictates that projects give some evidence of their 'broadly social, educative and/or emancipatory character' (Carter 1985).

The alternative enterprises, co-operatives and self-help projects in West Berlin fill the needs of 'political consumers' that are left unmet by capitalist producers. Socially-useful goods and services are aimed at this 'market segment'. This form of alternative marketing was not exhibited in the Technology Networks where socially-useful products were aimed at the public sector and areas of collective service provision, for example, aids for the disabled, and so on. The emphasis in Mehringhof on 'cultural commodities' rather than products serves to lower the costs of distribution since the projects are not involved in production and manufacturing.

Socially-Useful Products

The concentration on alternative products by the GLEB technology networks and similar initiatives in Britain grew out of concerns about rising unemployment and the possibilities of alternative production that were put forward in the Lucas Plan. Alternative products formed part of a 'pro-active' strategy in the fight against redundancies at Lucas Aerospace. The aim of the Lucas Plan was to demonstrate that there could be a redirection of resources from military to civilian sectors without causing job loss and economic dislocation. The term used to describe this strategy is 'arms conversion'.

The Lucas Plan stimulated a number of conversion plans in Europe and inspired enthusiasm for the notion of alternative production. The belief in the political efficacy of alternatives is derived from the view that trade union struggle for projects of this nature can point to alternative directions and provide a visible demonstration of another route of technological development.

In West Germany a number of Alternative Production Working Groups have been established in the larger military contractors including Blohm und Voss, AEG, MBB and Krupp. The German Metal Workers' Union, IG Metall, representing the largest group of defence workers, has been the main organiser of this activity. The Groups are concerned to explore the problems of employment insecurity in the defence sector, and several have developed alternative production proposals. These proposals vary from tyre-recycling equipment put forward by the Working Group in Voith in Bremen to much larger endeavours like the Combined Heat and Power System in Blohm und Voss. However, whilst product proposals are being put forward there is little opportunity for the development of product prototypes as organisations like the Technology Networks have, until recently, not existed in West Germany. The Innovation and Technical Advice Centres that have recently been established in Osnabrück and Bremen go some way towards rectifying this problem. The Centres are sponsored by trade unions, universities and local authorities; they are designed to address work-related issues such as new technology, product development and health and safety.

The plans for alternative production put forward in West Germany have tended to receive the same response that the Lucas Plan received from management, that is, the conversion proposals have been ignored unless particular products appear viable in terms of conventional market criteria. David Pelly (1985, p. 109), in an article on arms conversion, cites the example of the AEG Working Group:

> In AEG, for example, Working Groups were set up in a number of factories in response to the publication of management plans for closures and mass redundancies. AEG was heavily involved in manufacturing marine, power engineering and telecommunications systems. Over 70 per cent of the latter was for military application. The main proposals to come out of the Working Groups were a new local transport system for Berlin based on the renovation of the high speed railway (S-Bahn) and a 'block-heat energy' power station for Berlin's district heating system. The proposals would have provided work both in AEG and some other companies like Siemens through the manufacture of power supply, telecommunication and related equipment. But as these plans were being put forward AEG committed itself 'irrevocably' to its planned closures. Since management refused to consider the alternative production proposals, the implementation of these plans became dependent upon obtaining state support for the establishment of independent enterprises run by the workforce. This support was not forthcoming.

External constraints act to limit the possibilities of alternative production, as a person involved in the AEG project concluded: 'The most important step for the future of meaningful conversion projects must be that of considering the political steps necessary to help such internal factory initiatives and alternatives on the road to success' (Pelly 1985, p. 110).

Alternative plans as complementary to strategies of resistance have comprised a part of trade union strategies designed to deal with the introduction of new technologies. New Technology Agreements have been in operation in countries like Norway for some years. However the actual effectiveness of these Agreements has been open to question. Thus within the certain industries in Norway, for example, Borregaard Industries, the response to the ineffectiveness of New Technology Agreements has taken the form of 'blocking' the introduction of new technology systems until management agree to consider alternative proposals put forward by the trade unions. Noble (1983a) cites the example of the city of Bergen where city government workers have won a moratorium on the development of new technological systems until a long-term plan for technological change has been submitted by management; the moratorium has allowed unions and workers time to formulate their own 'Policy on Technological Change'.

Mike Cooley, a former member of the Lucas Aerospace Combine Shop Stewards' Committee and now the Director of the GLEB's Technology Division, noted that 'with the frantic drive forward of the new technology, we lack the time to examine the cultural, political, and social implications before infrastructures are established which will effectively preclude any examination of alternatives' (cited in Nobel 1983b, p. 80). During the time of the Lucas Campaign for the Alternative Plan a major acheivement was the security of a moratorium on the introduction of new technology which allowed the Combine Committee and CAITS to examine the implications of the introduction of new technology for the workers at Lucas Aerospace.

Thus the potency of alternative planning for socially-useful technologies is enhanced if there is the possibility of influencing the new technologies before they are introduced into enterprises. The concern of trade unions to challenge new technology and to participate in their design, development and use is exemplified by the International Association of Machinists of the USA. The IAM President William Winpisinger, said of their 'technology bill of rights' programme to a Congressional subcommittee: 'The objective . . . is not to block the new technology but to control its rate and the manner of its introduction, in order that it is adapted to labor's needs and serves people, rather than being servile to it or its victim' (Noble 1983b, p. 81).

In Sweden, Denmark and Norway the presentation of alternatives to the new technologies is an area in which researchers and unions are working together to present long-term plans for 'the office of the future' and 'the shop of the future', where technology is introduced to serve the needs of labour. Perhaps the most famous of these initiatives is in Denmark and Sweden, where workers and researchers are involved in the design of a worker-friendly computer-based printing system, the Utopia project.

The Utopia Project

The Utopia Project is a research effort funded largely from public sources including the Swedish Centre for Working Life, the Swedish Board for Technical Development, the Danish National Agency of Technology, the Nordic Graphic Workers' Union and the Royal Institute of Technology, Stockholm, and Aarhus University. This Scandinavian research and development project involves the trade union development of, and training in, computer technology and work organisation, especially text and image processing in the graphics industries.

The aims of the project are described by Pelle Ehn, the project leader, and Morten Kyng in a paper presented to the FAST Conference (1985 p. 8):

> For the last decade the ideas, work practices etc. of the first collective resource projects (action research projects studying the effects of new technology on work) have spread throughout Scandinavia. Local data agreements have been negotiated, data shop stewards appointed, and union clubs have formed their own investigative groups.
>
> But although growing, the extent and impact of these activities did not meet the initial expectations.
>
> • existing technology sets significant limits to the feasibility of finding alternative local solutions which are desirable from a trade union perspective.
>
> It seemed that one could only influence the introduction of the technology, the training, and the organisation of work to a certain degree. From a union-based perspective, important aspects like opportunity to further develop skills and increase influence on work organisation were limited by the available technology.
>
> Thus it was decided to try to supplement the existing elements of the collective resource strategy with union based efforts to develop new technology.

To try out the ideas in practice, the Utopia project was formed in co-operation between the Nordic Graphic Workers' Union and research institutions in Sweden and Denmark. The aim was to design computer support and professional education for integrated text and image processing.

We tried to summarize our experiences in a number of principles on which we wanted to base the design. These included:
—quality of work and products,
—democracy at work, and
—education for local development.

The project group consisted of graphic workers, computer scientists and social scientists involved in an alternative participative and skill-based design method, the main objective of which was to assist Unions in the translation of their social values, in the sense of job skill, quality of work and quality of the product, into new computer hardware and software for the printing industry. This process allows the possibility of workers influencing the shape of new technology before it reaches the shop floor. The collective resource strategy meant that knowledge and design were built up at all levels of activity, but with particular emphasis upon the local level. The accumulation of information on printing activities and existing work organisation formed the base from which designs were built. There was a great emphasis upon 'design by doing'. The design and conceptualisation stage was seen as crucial to the development of alternative forms of the technology.

Printing has usually been seen as a craft industry; however, over the past twenty years, the introduction of computerised text entry, typesetting and laying have had major effects on the print worker's role. New skills are required to operate the new hardware and software of computerized systems. In addition, the nature of the work has changed. For example, when lead was the material that was used printers made up an entire newspaper page in metal from sketches provided by the editorial department. Make-up workers could judge the quality of the page design because it was there in front of them. Lead was replaced by paper paste-up where page boards were still used. With the onset of computerization, page layout has become more abstract; early systems required workers to retain a mental picture of the page because the whole could not be displayed on the computer screen. Although this has now changed, make-up workers still experience difficulty judging page design because they do not work with the actual page or text. Malte Ericsson, a Swedish lithographer and participant in Utopia commented: 'It's almost as if you were working blind' (cited in Howard 1985, p. 45). Scandinavian print workers argue that the effect of the new computer technologies is a diminishment of both the quality of the product and the work experience.

The Utopia Project builds on the idea that the diminishment of the quality of the product and the work is not an inevitable consequence of computerisation. The automation of work could be replaced by computerisation which works to augment printers design skills, becoming, what Pelle Ehn terms 'an advanced tool for skilled graphics workers' (Howard 1985, p. 46).

From the beginning of the project it has found that the best method for discovering print workers' demands for new technology was by giving them direct access to its use. After some months, the Utopia Project acquired its own computer workstation programmed with a few sample layout functions which allowed the participants to refine their ideas for hardware, software and the organisation of work.

The move from Utopia as an abstract research project to an exercise in real technological development came in 1982 when Sven Holmberg, the president of Liber Systems, offereed Utopia a role in the development process of Liber. Liber was involved in a Nordic project to develop a fully-integrated text and image processing computer system known as TIPS; it would combine text entry, image enhancement, pagination, and layout in a single workstation. Participants in the Utopia project produced a set of 'applications specifications' that sought to balance technical capabilities with print workers' demands. For example, one of the print workers' demands was the ability to work with the whole newspaper page on the terminal screen—the screens are too small to allow this. The compromise developed by the Utopia team was software that provides make-up workers with 'lenses' that allow scale reduction, magnification or natural size. In this way make-up workers are able to change the text by working on only a portion of the page, or work on the overall design by reducing the page to fit the screen.

The 'specifications' also include aspects of organisation and training. However, the creation of new models for technology design and work organisation does not guarantee their use by employers. Liber launched its TIPS technology in 1984, and six of the systems have been installed including two at newspapers in Sweden and Finland. The Utopia recommendations for user-friendly software were incorporated into the system but most of the ideas concerning work organisation and training have been left to the discretion of employers, so that the impact of the technology will depend on negotiations between employers and unions.

The Utopia Project represents an attempt by unions to design their own technologies; however the manufacture of these systems involves compromises and the final mode of implementation depends on the outcome of employer–employee relations. Thus the problem of marketability raises its head again in a similar way to the problems that were encountered with some of the GLEB socially-useful product proposals. It appears that although the possibilities do exist for the design of alternative technologies that reflect worker and community needs, manufacturing and marketing of these products are areas that need to be carefully addressed.

The Community Memory Project

The demonstration of the viability of alternative designs and applications of new technologies requires the possibility of their manufacture and implementation. The Community Memory Project demonstrates the relative simplicity of the conceptual stage of a project and the difficulty of transforming conception into reality. Within this project the idea was that alternative applications of computer software could be financed by the commercial spinoffs of the research.

The Community Memory Project was based in Berkeley, California in the United States. It represented an attempt to create and deploy a non-hierarchical computer communications system. A member of the project team, Tom Athanasiou (1985, p. 38), describes the aims of the project:

> Community Memory is a system for the public management of public information. It is an open channel for community communications and information exchange, and a way for people with common interests to find each other. . . . All the information in the Community Memory is put in directly by the people who use the system: anyone can post messages, read any of the other communications that are there, and add comments or suggestions at any time.

In the early stages of the project, given that the time, energy and finance involved in a large software project would prove to be very great, it was decided that the underlying system software would be written in a modular form. This meant that the Community Memory system could be implemented as a generalised text/data handling 'toolkit' that would have commercial spinoffs.

A marketing company, Pacific Software, was created to which Community Memory would be able to licence its products, thus earning enough royalties to provide funding for the public programs. There were two major commercial spinoffs:

a) Sequitur, a sophisticated 'relational' database management system built upon

the Community Memory 'toolkit'. Designed to run on small computers, it is destinguished by the high degree of integration that it acheives between text and data processing.
b) X.Dot, a portable C-language implementation of an international standard data-communications protocol, X25 (Athanasiou 1985, p. 43).

Problems were encountered in dealing with the commercial and market pressures. The informal nature of Pacific Software, and the lack of managerial mechanisms, led to difficulties of proper planning and scheduling, so that the time spent on code changes and other features that the market demanded dominated the activities of the project team, to the detriment of Community Memory. Pacific Software went bankrupt.

The realities of the market impinged on the project team in a political sense also. The X.Dot system was launched into the telecommunications market by Pacific Software, but the company was approached by a firm in Johannesburg who were building an airline reservation system and needed an X.25. The sale was refused and subsequent sales to South Africa were prohibited. Sales were also restricted to the military, through the limitation of Pacific Software sublicensing rights to 'commercial and non-proprietary' applications. Tom Athanasiou (1985, p. 46) explained the problem:

> I, along with others, opposed the sale on the grounds of its unique benefit to South Africa. X.Dot, designed to be easily transportable among a variety of machines, would have been especially useful to South Africa, suffering as it does a technology boycott of some significance. This same criterion of uniqueness led me to support sales of Sequitur to the military, when not much later they became an issue.
>
> It was the agreement over these sales that first emerged as open conflict. In the absence of the clear cut ethical/political imperative provided by a pre-existing boycott unanimity broke down. There was reason to doubt, whatever sales we denied ourselves, that our denial would have an effect. Sequitur was of no unique benefit to anyone: there were dozens of systems that could do the job equally well (perhaps better). Furthermore there was a strong feeling that, whatever military sales policy we were eventually to adopt, it would be necessary to keep it quiet. The logic of the market, many felt, dictated that we do not endanger Pacific Software by making it appear political or constrained in undefined ways.

In this way the value of the project had to be weighed against the concrete political and economic realities of the market.

Discussion

While the GLC/GLEB experiments have been an inspiration to many of the projects just described, not all of these flowed from them. Indeed some pre-dated them—and some were developed in parallel. But either way, there does seem to have been a confluence of ideas concerning the role of socially-directed technological innovation.

The examples looked at have also all been at the local or regional level—and not part of a national network of schemes. However the British Labour Party has taken a strong interest in this approach and has included proposals for a national network of 'Local Enterprise Boards' modelled on GLEB in its programme—as outlined in its 1986 *Charter for Local Enterprise*.

The Labour National Executive Committee's statement to the 1986 Labour Party Conference (NEC 1986, p. 6) spelt out their proposals as follows: Labour will work through enterprise boards and other agencies and where appropriate, directly through local authorities, to support industrial development in a variety of new and innovative ways. They will be encouraged to:

- Provide funds for investment in return for equity shares. In this way Local and Regional Enterprise Boards will be able to take an active role in shaping the development of local industry instead of acting merely as passive money lenders.
- Extend their support beyond finance to include business advice, assistance with finding property, funding of R & D, and training.
- Support further expansion of workers' co-operatives—where workers have a genuine say over the strategy the company pursues and the way it is run.
- Encourage the development of community industry by providing finance, premises and management services.
- Develop research links with local polytechnics and universities and set up 'tech-nets' (information exchange systems for new products and process technology).
- Provide financial and technical assistance to people wishing to go into business but usually not viewed with sympathy by banks and other financial institutions—notably, redundant workers, women and ethnic minorities.

It also proposed the establishment on a national public agency 'British Enterprise' which would: 'Establish operating arms able to achieve the comprehensive restructuring of specific industrial sectors or to develop particular industrial processes—for example, in the field of health' (NEC 1986). A crucial task for British Enterprise will be to support and boost new technological industries. It will take equity stakes in new and existing firms. This will enable a Labour government to achieve a high-technology industrial base specialising in advanced manufacturing technology applied to new and traditional products and processes. It will work with individual companies to capitalise on their potential and existing technological strengths. British Enterprise will be ready to work jointly with the universities and entrepreneurs in the vital task of transferring research and development from the research labs into production for the market-place. British Enterprise will thus act as a spur for the development of new technology and its application to new and established industries.

In British Enterprise-assisted firms, management and workers—through their unions—will draw up business plans, specifying the company's long-term strategy, and covering areas such as investment, research and development, marketing strategy and training. British Enterprise will monitor the progress of these business plans—taking full account of the views of those who work in the enterprise—and will thereby be able to identify the specific needs of particular firms, providing packages of support geared to their detailed needs.

British Enterprise will have the financial backing it needs to fulfil its role—through finance from government, the British Investment Bank, and by being able to raise finance from the market. In determining the sectors and processes in which British Enterprise intervene, it will not fight shy of tackling new sectors of the economy in which government has not been involved before. By introducing new forms of government equity support, BE will be taking a stake in British industry, enabling social ownership to be extended throughout the economy. Thus GLC/GLEB ideas could be taken up on a fairly major scale.

Conclusion

The GLC and GLEB model for industrial development and employment has been greeted with enthusiasm by countries in Europe and elsewhere which are facing the same problems of economic decline. The lines that can be established between enterprise boards, and technology exchange agreements between different regions in Europe look to the production of appropriate technologies to enterprises and the access to different markets. The links with the Third World demonstrate the possibility of trade agreements that can prove of mutual benefit to partnership countries. Product diversification within enterprises and new product innovations that are socially useful in some form, has been seen as the best way to connect technologies with needs, particularly the creation of employment.

The creation of employment in competitive market economies is dependent on the commercial viability of new products. The Technology Networks and similar initiatives in Britain have attempted to 'market' the idea of socially-useful products by the encouragement of user development in product design. A major factor has been the establishment of academic and community links for the purpose of technology transfer of ideas within polytechnics, to manufacturable products in the community. The links between science and the community, whether through products as in Britain or by way of scientific and technical information, as in the Science Shops in Europe, have proved to be not without difficulties, but represent steps towards a new approach. The European Science Shops operate in a different way to the Technology Networks so do not encounter the problems of manufacture and marketing of socially-useful products. However, it is possible to view their ideas in a similar way, in the sense that community needs revolve around necessary information, for example, much of LEEN's work is aimed at 'marketing' socially-useful information on energy-related issues, as well as socially-useful products.

The alternative marketing style of the West German Mehringhof collective focuses on socially-useful information in the sense of retailing and distribution of goods and services which represent a particular political perspective. The 'market segment' of politically conscious minority groups comprises the consumer base. The aim of socially-useful production as the production of social relations is to change attitudes and consumer practices. The change of the attitudes of consumers to socially-useful products is a major factor for the Technology Networks if products are to become marketable. The distribution of both information and products is central to strategies designed to develop alternative routes of technological change.

Product design which embodies different social relations, that is, user involvement, formed the base of the socially-useful products that were put forward in alternative corporate plans as initiated by the Lucas workers' *Alternative Corporate Plan*. Alternative planning has been seen as a positive response in the areas of arms conversion; the transference of products from military sectors to civilian sectors, and new technology, for workers seeking to influence the direction of technological change.

The problems encountered by the Lucas Plan product proposals, that is, that they were ignored by management unless they appeared viable in terms of conventional market criteria, were shared by workers' groups in West Germany. The development of product prototypes is seen as an important element of the strategy; this area is addressed by the Technology Networks in Britain and the Innovation Centres recently established in West Germany. However, the manufacture of product prototypes has proved to be difficult. The GLEB have attempted to address this area through their technology exchange programme, but additional outside finance is required to support the manufacturing costs of products; see, for example, the Airlec case study. An alternative to this is exemplified by the Utopia Project, where a large manufacturer in Sweden incorporated the print workers' specifications into the design of the TIPS technology. However, the form of implementation of the technology will result from the outcome of negotiations between employers and workers. A third way of securing the manufacture of alternative technologies or the application of alternative technologies is exemplified by the Community Memory Project where the idea was to use the commercial spinoffs via the design of the technology to fund its alternative applications. Although a few terminals are now in operation, the project was greatly affected by the political and economic realities of the market.

This resumé of some of the similar initiatives in Europe and elsewhere designed to influence and develop alternative technologies, whether through the dissemination of information or the production of product prototypes, suggests two major conclusions. Firstly, that technology, if it is to fulfil needs, requires manufacturing and production opportunities, which suggests commercial market viability. Secondly, that the manufacture of socially useful products presents a major problem of finance, access to appropriate manufacturers and is constrained by the realities of the market.

Further issues raised—particularly by the British Labour Party's proposals for national-level intervention agencies, is that of political constraints. The GLC and the GLEB came up against severe constraints imposed by a hostile central government—national-level initiatives, even ones established by a Labour government, would also be likely to face problems.

In Part Three analysis of the GLEB experiment and other initiatives is made and consideration given to the implications of this for the development of alternative technologies.

PART III: Analysis and Implications

Introduction

The GLEB innovation initiative was based on two beliefs: first, that it is possible to direct the innovation process towards social objectives; and second, that the market mechanism is inadequate to ensure the satisfaction of social needs. It was envisaged that the establishment of an 'alternative' technological paradigm based on the concept of 'social use' and user involvement in product design and development would act as a framework for a trajectory of socially-useful product development. In contrast to the market-led approach, need-led innovation initiatives were orientated towards the public sector provision of services in the areas of energy, aids for the handicapped, transport needs of disadvantaged groups, and so on, the public sector providing the demand factors for technological change.

Socially-useful product development and production for social need as guiding principles distinguished the GLEB approach from more conventional innovation support initiatives. But in practice, production for social need is not fully achievable outside a socialist society. The curbs on public spending at national level in the UK have served to preclude the possibility of the GLEB realising these objectives, apart from in an 'exemplary' way. For example, in some cases the networks have been able to indicate how this model would operate; an audio system for schools was developed in the networks, and manufactured at the GLEB-supported London Production Centre. The Local Education Authority acted as the public-sector market.

The aim of the GLEB innovation strategy was to address the problems of unemployment and wasted resources, such as empty factories and redundant machinery. In a competitive market economy, employment creation requires the economic viability of new product innovations if enterprises are to survive and jobs are to be secure. Thus to a large extent the problems with the GLEB innovation approach lay in the attempt to operate within a competitive market economy but with production geared to the provision of needs, rather than at a particular market segment or niche of the consumer market.

There is, at least in principle, a false dichotomy in setting socially-useful products against products manufactured for individual consumption and sold through the market; most technologies exhibit some form of use value (Bodington *et al.* 1986). The point about socially-useful products is that the demand is assumed; because the need exists and relates to social service provision then the product should be developed and manufactured to fill this need, thus production for social need is politically prescriptive. However, while the GLC did have some control over its own purchasing power (for equipment and services), the allocation of resources to the public sector for spending on social services is a national political problem and outside the influence of the local authority. In terms of the innovation process the GLEB were able to address technology-push factors but were unable to influence public sector or market demands.

In this section the GLEB case study and other attempts to influence innovative activity are discussed in relation to the factors identified in Part One as influential for the selection of routes for technological development. Chapter 5 focuses on the GLEB initiative for innovation support and user involvement in product design. Chapter 6 looks at the wider economic and political context of the GLC and the GLEB.

5 Analysis

Innovation

Public sector intervention in the innovation process by the GLEB focused on the provision of 'seed-corn' finance for selected product innovations that included a range of technologies, that is, new technological development, commercially-orientated product innovations and socially-useful products. The technology networks were established to provide the R and D facilities for product innovations. In particular, three points were emphasised that sought to distinguish the GLEB initiative from other forms of innovation support: systematic user involvement in product design, employment creation and the linkage between technological opportunity and unmet need in the form of socially-useful products.

There were three main areas of innovation support. First, the GLEB attempted to support 'in-house' innovation in existing firms that were in receipt of GLEB assistance. Here the support for product diversification was seen as central for the economic regeneration of failing enterprises, that is, enterprises that needed to increase their competitiveness in a particular market. The GLEB provided risk capital for several small, often new, enterprises based on new products together with several large existing companies. In some cases the attempt was made to support the 'clustering' of a number of similar firms to establish a more viable innovation package; this was the strategy with AMD and the Synergy Centre. In principle, the technology networks could feed in new product ideas, but in practice, innovation was 'in-house' and mostly focused on new microelectronics-based technology.

Second, product innovation support was geared towards product ideas that could exhibit characteristics of social use and also prove commercially viable. For example the Playbox developed in the London Innovation Network (LIN) met the need for a toybox to be used in creches and nurseries that did not take up too much space, and efforts were also made to aim this product at the consumer market.

Third, the GLEB supported the development of socially-useful products which, although needed, were unlikely to prove commercially viable, for example, aids for the disabled. These products were developed in the various networks in collaboration with community and user organisations. It was hoped that orders for the equipment would be forthcoming via the provisions offered by the Department of Health and Social Security for people with special needs, or via local council purchasing. However the cutbacks in local and national welfare provisions have made it much harder to obtain this sort of backing.

The diffusion of existing technologies and the commercial development of new products is a qualitatively different route to technological development than socially-useful products. The latter is much harder because it involves political processes aimed at the establishment of forms of exchange that are based on social accounting; that is, perceived economic importance is secondary to the fulfilment of a social need.

Given the financial and political constraints, the GLEB were faced with two choices: either to support commercially-orientated product development which, though problematic, offered the possibility of some commercial success, and thereby some employment preservation, or to concentrate on the development of alternative technologies which meet social needs but are unlikely to be developed commercially. In general, the GLEB chose the first route and attempted to influence the application and diffusion of existing technologies. The socially-useful design and development of products in the Technology Networks served to exemplify the possibility of alternative directions for technological development but was increasingly marginalised.

With the high hopes for the 'radical' experiments of introducing an alternative technological paradigm, and thereby instituting a trajectory of socially-useful product development, the question emerges as to why this was the case.

Technology-Push

In the economic approach the analysis of innovation is based on the assessment and evaluation of new products and processes according to conventional market indicators of 'success' and 'failure'. However, this form of assessment is not wholly appropriate for the GLEB experiment. In the first place, the GLEB has only been in operation since 1982 so that the time-scale for this form of assessment is too short. The innovation process in many cases is long-term and the evaluation of the effectiveness of product and process innovations can only be made with hindsight. In the second place, a key feature of the GLEB approach was the emphasis on the social context and social process of innovation, that is, user involvement in product design and development. A more informative way of looking at the GLEB experiment is to identify the factors that mitigate against the attempt to support alternative forms of product development and innovation.

Political and economic factors influence the selection and development of technological trajectories. Priorities for investment in R and D, and market demand factors respectively, are likely to preclude the possibility of introducing alternative technical solutions to user problems.

For a product idea or invention to become an innovation, technology-push must match with market-pull. The GLEB were able to address technology-push factors in terms of risk capital and R and D support.

One of the major problems for inventors is access to finance. Few investors are keen to provide risk capital for longer-term ventures; the time-scale from idea through prototype development to manufacturable product can prove too long. Until investment is considered to be necessary or potentially profitable, technologies will not be developed. In consequence, potentially new ideas and markets may be ignored, and more generally, the concern with commercial viability may result in the neglect of projects which have high social value potential. Investment in projects in their early stages of development was seen by the GLEB as a necessary area to address if intervention in the innovation process was to be strategically effective. In this way the GLEB could support selected technologies which were considered appropriate and desirable.

In its early stages of operation the GLEB found that enterprises in need of support did not require much technological aid, rather it was more a question of the retention or expansion of existing production activities and markets in selected sectors. The lack of statutory powers of intervention for local councils meant that the GLEB could only intervene in companies that approached them. In effect companies in financial difficulties. These difficulties could be attributed to a variety of factors, for example bad management or outdated equipment, rather than the need for assistance with the development of new products. However, with the management of demand outside the influence of local authorities, help with aspects of production was considered the most appropriate means of economic regeneration. Innovation support in this context requires fitting into an enterprises existing trajectory of product development; the room for the introduction of new or qualitatively different products is minimal.

New product development is easier in 'start-up' enterprises, but there are the attendant difficulties of small firms operating on the market. The GLEB, via the Technology Networks, has been able to stimulate the development of a number of selected innovative products by supporting the work of individual inventors: Airlec and Pedelec were mentioned in the case study. This area of innovative activity, product development to prototype, does not incur huge costs so that it is seen as appropriate for assistance given the financial constraints on local authority spending on economic development. However, although products have been developed to marketable product stage, industrial partners and private investors have not been easy to find for the stage of moving to full-scale production. John Palmer, the director of the GLEB's Information Division sees this as one of the problems faced by local enterprise boards: 'There have been cases where local enterprise boards have facilitated the development of innovative and potentially world-beating new products only to find that it is next to impossible to secure commercial exploitation' (*Guardian* 24 December 1985).

So although there were confident assessments concerning the commercial attractiveness of these products and orders had been placed on the assumption that manufacturing partners could be found, both developments subsequently had to be halted. This was despite clear indications of commercial viability, for example, interest in the Pedelec was particularly encouraging both from within the UK and overseas, with the Chinese government indicating an interest in 10,000 units.

Whilst the GLEB were able to provide 'seed-corn' finance to inventors to support projects at their development stage to marketable product, they did not have the resources to invest in full-scale production and manufacturing. The original idea was that such products would prove attractive to commercial interests, that is, once developed to manufacturable stage, the product would be taken up by the conventional market mechanism with possibly the original company itself being taken over by a larger one. But this did not happen with either the Pedelec or Airlec. There are perhaps several reasons to account for this: first, small firms manufacturing single products are not very viable commercial projects; second, the costs involved for enterprises moving producting processes and products into a different sector are often prohibitive; and third, it is often difficult to obtain finance for products that are not within existing areas of development, for example, it is easier to obtain finance for products exhibiting 'hi-tech' characteristics than for products that are considered 'low technology' or in some way novel. Pedelec and to some extent Airlec failed on all three counts: new products, new firms and new markets.

Innovation and the Technology Networks

Innovation is necessarily a long-term process in that in some cases the time-lapse between an invention and its eventual application and use can be ten to twenty years or longer, depending on factors such as market demand and financial investment. The creation of the Technology Networks within the political framework of a local authority meant that the short-term political priorities of demonstrating the effectiveness of the strategy, in practical terms, that is, actual products and employment creation, were in conflict with the longer-term process of innovation. Thus, in many cases the choice of product development in the Networks was conditioned by what could be produced in the short term; for example, the energy network, LEEN, in its 'Jobs for Warmth' programme developed a number of short-term innovations that could be balanced by long-term projects such as Combined Heat and Power. Given the short time-scale and limited resources the Networks were concerned with incremental innovative steps that could demonstrate the alternative applications of existing technologies; for example, LNTN were very involved in the alternativc applications of computing networks for community organisations and community health care.

The Networks were established, in part, to link the technical resources of polytechnics to community needs. The emphasis on the technical aspects of innovation support led, initially, to a focus on product development. However, the process of the stimulation and development of product innovations has proved to involve a number of problems, and is slow. Therc was no shortage of ideas, the computer product bank created by ITDG held a list of new product lines but it was not easy to find people to develop the ideas, despite the technical resources of the Network workshops or to finance the projects. Although the GLEB were able to take up some of the ideas the time-scales involved in developing a prototype product from an idea were long and in conflict with the aim of immediate job creation.

The 'Jobs from Warmth' campaign, for example, grew out of the concern about the high cost and poor operation of the heating systems in council-owned flats. Technical options were reviewed including the idea of using a medium-sized heat pump to supply district heating networks. CAITS took a leading role in this activity. The notion was to find a local manufacturing enterprise that would expand to produce such units for several housing estates. There were plans for a demonstration unit; a rotary engine gas heat pump concept originally as part of the Lucas Aerospace Workers' Campaign was favoured, to be set up on one estate with possible funding from the local council. However, the project was halted due to insufficient finance, while LEEN continued with development work on the heat pump, the possibility of manufacture and implementation remained remote. The main point to be gained from this example is that given LEEN's technical resources, there was no real problem with the technology but rather the problem lay with the political commitment of resources for implementation.

Need-led innovation is thus a political rather than a technological problem. The recognition of this led to the development of the other side of the Network strategy; campaigning. For LEEN this involved energy advice and support work, with the emphasis on relatively simple energy conservation measures. The argument was that jobs could be created more rapidly and needs met more effectively by focusing on the provision of advice and support on energy efficiency. In the longer term new products would emerge from this process if they were considered necessary, for example, new energy-efficiency monitoring equipment.

This strategy exhibits an approach which is more design-orientated whereby intervention begins with the problem area and technical needs and solutions are defined more clearly, that is, a problem-led perspective. Working in this way LEEN have achieved some success in building alliances with local community groups and local councils to provide some user feedback about needs. The identification of needs and the provision of advice and monitoring services has led to the development of a number of new product ideas that are being considered for production at the London Production Centre, the GLEB-supported Technology Centre.

Attempts to introduce innovation initiatives highlighted the problem not of developing new technologies but rather of the implementation of technologies that already exist. Production for social needs requires increased public spending in the area of collective service provision; this is effectively limited by the economic environment. The recognition of this meant looking at the possibility of the development of socially useful products aimed at individual purchase and consumption via the private market mechanism. The attempt was made to produce products that consumers would choose to buy because of their use-value. The brief of employment creation was particularly dependent on this arm of activities, at least in the short term. Thus the Networks attempted to link social use characteristics of products with commercial viability.

Of all the Technology Networks established, the London Innovation Network (LIN) was perhaps the most concerned to address the problems of manufacture, marketing and the commercial viability of socially useful products, with products such as the Whitfield Technology Bench (a technology bench for schools, designed at the Design Development Unit, and aimed at the educational furniture market) and the Playbox, mentioned earlier. As with the larger GLEB initiatives it was found that, whilst product development to prototype is possible with relatively small sums of money, access to capital is the greatest problem for manufacturing. The GLEB have addressed this problem in the areas of high technology in the Technology Centres, but for the Networks concerned with the support of local employment in some so-called 'low-technology' sectors access to manufacturing capability remained a problem; there are only twenty-four manufacturing co-operatives in London and many are ill-equipped and undercapitalised. For the manufacture of the Whitfield Technology Bench, it was necessary to look outside London for manufacturing capability. The links with the commercial manufacturing world presented the problem of the role of the Networks as R and D support to the private sector.

The initial basis of the creation of the networks as facilities for technical support for product development aimed at the public sector market; production for social need led to a lack of attention to marketing, commercialisation and implementation. The rhetoric of 'alternatives to the private market mechanism' allowed attention to finance and distribution to be neglected. This is most markedly evident in the recruitment of staff for the networks; most employees have technical skills but do not have entrepreneurial skills. To an extent this can be accounted for, particularly in the area of marketing, by the difficulty of staff recruitment in an area where there is not a ready pool of people with the right types of skill to achieve a balance between social use characteristics of the product and commercial marketability.

Product design and alternative product design occupy a central place in the innovation process, however the market or need forms an integral part of the product design, that is, products are not designed and produced and then subsequently marketed and distributed. The push-me-pull-you relationship between technological capability and the market is central to innovative activity.

The commercial viability of socially-useful products remains a problem for the networks, for the nature of the wider competitive market environment constrains the possibility of creating and developing products in an alternative way.

Technology Centres

The third form of 'in-house' innovations supported by the GLEB was in the area of 'new' technology enterprises via the Technology Centres. The aim was to foster the development of 'high-technology' manufacturing centres by clustering a number of product design and marketing companies around a large existing enterprise that would act as manufacturer, for example, the London Production Centre discussed in the case study. In this way, the problem of gaining access to manufacturing opportunities and moving from product development into volume production was addressed. The similarity between the smaller enterprises and the large firm in terms of product range means that the manufacturing of these products fitted into the existing trajectory of the larger enterprise.

The notion of stimulating innovation by building on a firm's existing trajectory is also evident in the Technology Exchange Programme established by the GLEB in 1985. The Exchange acts to reinforce the approach adopted in the Technology Centres. The focus has tended to be in the area of electronics, the prime concern is to match particular product innovations to the product facilities and market experience of specific companies. The potential for success in market terms for this strategy is exemplified by the Technology Centres. Several new ideas in the electronics field, along with several 'start up' firms established initially within the London New Technology Network have transferred to the London Production Centre site, and been taken up by the existing company there, Broomhill Electronics. The design and marketing functions remain with the smaller enterprises whilst manufacturing functions are transferred to the larger enterprise.

The first of the satellite companies has produced an audio system for schools with the purchasing power of the Inner London Education Authority (ILEA) providing an initial market. The aim is to set up another company to produce and market the microelectronics-based energy saving controllers for industrial and domestic heating and ventilation that have been developed in the energy network, LEEN.

Within this approach the emphasis is not on the creation of small start-up companies in isolation but rather to link the innovative power of the Networks to a federated cluster of development, production and marketing companies with adequate manufacturing capacity.

The Technology Centres and Technology Exchange exhibit the features of a conventional approach to innovation in that the products are market-oriented and largely in the areas of rapid technological development, 'hi-tech' industries. Thus, the tendency to technological convergence appears difficult to resist.

The potential success, in market terms, of this perspective to the stimulation of innovation suggests that the route to product innovation is fairly well defined by the economic environment. Firms exhibit a trajectory of product development which is closely tied to economically-defined needs expressed through the market. Technical change comprises a series of innovative steps arising from indicators signalled through the interaction between market needs and technical possibility. The selection of goods on offer tends to fall into similar trajectories of development as companies seek to imitate market leaders.

Discussion

In the GLEB innovation strategy, the stress was on the sphere of production. New product innovations were seen as the means by which enterprises would achieve economic viability. This emphasis on production rather than distribution and marketing meant that the approach was essentially supply-side or technology-push. The ability of local authorities to influence market forces and a changing economic environment is minimal. For this reason the provision of financial aid and technical expertise appeared to be the most sensible route to innovation support.

However, market constraints make the development of alternative technologies difficult. Whilst it is possible to support these technologies from the supply side with relatively small investments, product development and innovation is closely tied to changing 'market demands'. Reliance on technology-push factors tends to underrate the significance of economic factors that are crucial if an invention is to become an innovation.

The argument that innovation is influenced by the recognition of a 'need' which producers seek to fill has to be reinforced by the understanding that these 'needs' are economically significant market demands. Needs backed by purchasing power comprise the link between technological innovation and the market. The effectiveness of social needs as the basis for the stimulation of innovative activity is marginal in competitive market economies, that is, unless these needs are also economically significant. The GLEB was unable to influence demand factors from the public sector so that the mechanism did not exist whereby minority groups without economic and political resources could influence the innovation process.

That said, the GLEB was able to create employment—over 3,000 jobs[1]—by stimulating some novel, exemplary projects, in which participation and user involvement were significant features.

The Social Organisation of the Innovation Process

The GLEB was to provide facilities for 'non-experts' to influence the product design and development process. Design plays a central role in the innovation process. The technical solutions that emerge in relation to user problems exhibit a number of potential design configurations that are dependent on the priority given to certain criteria, for example, the different designs of a product that will emerge if priority is given to cost or alternatively to materials. Socially-useful design is based on the direct involvement of the user or 'design-by-doing'. The Technology Networks were established to provide the facilities, first, for participation in product design, and second, to assist with technical expertise by linking the resources in local polytechnics to the community.

If the relationship between the polytechnics and the Networks is looked at first, how did this work out? The first point to be made is that the establishment of links and joint working arrangements was variable; for example, Thames Technet had very close links with Thames Polytechnic, whilst LEEN's relationship with the Polytechnic of Central London (PCL) and the Polytechnic of the South Bank were more dependent on the commitment of particular individuals. Some individuals have been drawn closely into LEEN's work via the PCL-linked Energy Conservation and Solar Centre (ECSC). Links were forged more easily by working on a particular problem that was relevant both to the polytechnic and the networks, rather than on more general issues. In some cases the establishment of a unit in a polytechnic that was a part of the Network was more useful, for example, the Design Development Unit at Middlesex Polytechnic was attached to the London Innovation Network. The GLEB provided funds for work on student projects that were directly relevant to the network. Links were established with, for example, engineering students at Imperial College, and with marketing students at Thames Polytechnic.

One difficulty for the networks of collaboration with the polytechnics was that the polytechnics were suffering from drastic financial cutbacks, with extensive staff reductions, loss of morale and staff working under immense pressure and uncertainty. Consequently, they tended to look to GLEB as a possible source of 'research' funding, rather than as a partner in developing ambitious new 'social' projects. Much as they might have preferred to allocate resources to the development of socially-useful products, their main concern was their fight for survival. This is not to imply that those individuals and groups from the polytechnics involved in the Networks were not supportive. On the contrary, they were very keen to develop the ideas they had and work hard in setting up the Networks.

Interestingly, by contrast some of the polytechnics outside of London, namely Sheffield City Polytechnic and Lanchester Polytechnic in Coventry, saw the development of local council-supported employment-creating 'new product centres' (like SCEPTRE and UDAP) as part of their survival strategy; possibly this reflected the different political environment in these cities.

Another form of academic and community links exhibited by the networks was between the full-time staff and local community organisations. In general, the staff of the networks were from academic and technical backgrounds—with only a small number of personnel recruited with non-technical backgrounds or with experience in the voluntary sector.

As with the Science Shops in Europe, it had been hoped that technical organisations such as the networks would act to reduce community deference to scientific and technical expertise. Certainly, the emphasis that emerged on campaigning as part of the networks' brief necessitated a bias toward political mobilisation around technology issues. But links with the community were not always easy to build. Whilst Thames Technet attempted to address this problem, the difficulty of finding people with skills in both technical areas and community mobilisation was recognised.

Initially, the polytechnics were seen to be the source of product ideas; however, the move towards a problem-led approach, as discussed in relation to LEEN, meant that product ideas were developed out of the social context of implementation. The technical resources within the polytechnics were tapped to provide back-up support if it was deemed necessary.

User Involvement in Product Development

The Technology Networks were created to provide the facilities whereby producers and users could develop an alternative approach to product design which integrated the skills of those involved in the design process. The aim was to discover ways of organising design and production so that socially-useful and human-centred products could be developed.

Design involves technical knowledge but, just as importantly, it is also based on experience and observation; for example, in the Utopia Project 'design by doing' was emphasised whereby ideas for changes and improvements to products and processes grew out of the experience of working with the hardware.

The concern of the Technology Networks was how to involve users in the design process, and secondly, how to encourage the articulation of needs. It is unusual for users to be able to pinpoint the nature of their dissatisfaction, thus the emphasis in the Networks was on the establishment of links with community organisations that could begin discussion on the nature of needs, for example, the LIN project on Disablement and the Domestic Environment. Obviously, it takes a long time to build links and the danger with the product-orientated networks particularly was that products would be based around existing technical capability rather than user needs. The LEEN found that this was a tendency in the initial stages.

Design can be perceived as a 'solution-focused' activity that is based on the identification of a problem but the difficulty with the encouragement of user involvement in product design is the ability of non-designers to being to generate new possibilities. The remoteness of technology from everyday life in the sense of the lack of participation in technological development that is engendered by the privatisation and specialisation of the innovation process and the increasing technical complexity of many technologies results in the tendency to reinforce attitudes of deference to technical and design expertise. The Networks attempted to provide a bridge between technology and the community; this means adopting a pro-active approach to participation. The people most likely to approach the Networks were single inventors rather than groups of users or producers; in part this can be explained by the fact that the context of the local authority is qualitatively different from the experience at Lucas Aerospace (which was influential in the establishment of the Networks) where alternative plans for socially-useful products integrated the skills of manual and professional workers, in a context of struggle against management plans for rationalisation and redundancy. Thus, the Lucas workers were closely integrated and facing a common enemy, and the concept of 'socially-useful' products gave the rallying cry for the campaign.

The innovation process envisaged for the Networks was based at least on the idea that requests for advice, ideas and help could come from outside—from groups of workers threatened with plant closure via GLEB, or community groups contacted by the GLC's Popular Planning

Unit. In the event, fewer of these emerged than was expected: certainly not many needed 'new products'. The networks therefore had to be pro-active (taking the initiative) rather than reactive (responding to requests for help). The problem there was liasion with the GLC and GLEB 'sector strategies'. In principle, the sector plans developed as part of the GLC's London Industrial Strategy and GLEB's overall sector investment plans ought to link to and provide direction for the Network's activities. But co-ordination was not always achieved successfully—partly because of external factors.

The GLC was, after all, under concerted attack and fighting for survival: longer-term 'innovation' projects (like the networks) were therefore less likely to attract as high priorities as projects which might deliver visible results quickly (for example, by investing in centres based on *existing* companies like AMD/the Synergy Centre).

So at times some of the subgroups within the GLC/GLEB were pulling in different directions, some emphasising 'new products' and innovation, some investing in property and other sector investment, all with a different time-scale.

A more co-ordinated 'project team' approach with representatives from each relevant group liaising on each project was subsequently introduced in an attempt to deal with this type of problem. But even so, it is obviously hard to direct *invention* towards specific sector strategy defined goals. GLEB and the technology networks were able to *select* what they felt might be commercially viable, socially and environmentally appropriate and sectorially relevant products from the various ideas on offer, developed more or less spontaneously by an inventor (for example, the Pedelec). And they could choose to fund promising individuals, groups and lines of development. But they could hardly expect to totally steer the actual 'creative' invention process itself. A certain amount of serendipity and autonomy is obviously vital.

As has become apparent, steering the innovation process 'from the front end' involves a complex process of choice, stimulation and co-ordination—balancing protection of spontaneous creativity against strategic needs and time and resource constraints.

Implementation and Management

Co-ordination problems like this were arguably, however, only part of the problem. It could be argued that there was at least an initial basic weakness in the overall innovation concept—a lack of attention to the final stage in marketing, commercialisation and implementation. The original concept was that GLEB would assist companies in trouble, or with ideas for expansion, with funds and technical expertise. Subsequently the companies would deal with the marketing and commercialisation of existing or new products. GLEB did not wish to get involved with that directly: it did not have the expertise. But that proved to be the major problem for some of the smaller companies, as was the availability of appropriately qualified managers in general. As Palmer put in in a review of the problems faced by local enterprise boards (*Guardian* 24 December 1985) there was a

> quite appalling dearth of quality professional management to be found, particularly in small to medium sized businesses. All too many firms who find themselves seeking help from local enterprise boards or other agencies are in that position because of the low grade quality of their management as much as any other factor.
>
> Even when enterprise boards start off with a determination to maintain a 'hands' distance' relationship with their investments they often find themselves being drawn into their direct management.

Obviously this would put even more strain on the very limited resources of organisations like GLEB—which after all was set up as an investment and innovation support agency, not a management consultancy.

With GLEB finding it increasingly necessary to adopt a 'hands on' approach and develop its capacity to provide managerial and commercial advice, its ability to support longer-term innovation activities would obviously be impaired.

Enterprise Planning and Participation

Commercial success was of course not the only aim of the GLEB approach to investment and innovation. Equally there was a concern for improving working conditions and industrial democracy. Indeed, it could be argued that setting up a new product innovation system would, in the end, be a less important contribution than stimulating new forms of co-operative management and worker participation. Certainly, GLEB pushed hard to ensure that work-force involvement was achieved via the Enterprise Plans negotiated with each company. It also paid considerable attention to health and safety conditions and the actual labour process, for example, pushing for the introduction, where possible, of skill retraining and enhancing 'human-centred' production systems. Thus, socially-directed innovation concerning the organisation of production (that is, process innovation) was seen as equally important as product innovation (a point which was made strongly in the original Lucas Plan).

However, success in achieving worker influence over such matters clearly depended on there being strong work-force representation via trade unions. But in some of the companies that approached GLEB, trade union organisation was weak or even absent. In some cases it proved hard for workers to participate fully; often they did not have the confidence.

The same problem occurred in relation to more general community participation; for one thing building up a local capacity to participate effectively takes time and that was not available. The GLC's Popular Planning Unit was established to stimulate some local debate on specific needs and problems in general; this was patchy and did not lead to much involvement in the GLEB innovation programme.

That said, some of the Networks have been successful in building local links and in working for and with local community groups and user groups and organisations on specific projects and plans (for example, local energy plans).

While direct community participation in the innovation process has been patchy, the GLC's various consultative exercises and the 'local plans' developed with the help of the Popular Planning Unit have at least provided an input into the GLC/GLEB decision-making system—identifying specific problems, needs and areas of likely demands.

Assessment

In what way was the GLEB's approach different, in practice, from a conventional approach to investment, and, more specifically, was it able to stimulate and direct the innovation process?

Socially-directed innovation involves three main factors: user involvement in product design, social use characteristics of products and a direct link with employment. The theme of production for social need included the innovation initiatives in the networks; again because of the short-time scale it is not possible to assess innovation in terms of product development although some products have emerged from the networks, rather assessment is made on the ability to realise their objectives in terms of the social process of innovation of socially-useful products.

As the discussion of the initiative indicates, the time factor has proved to be of major importance. Academic and community links and user involvement in product development require a fairly lengthy process of discussion and liason before links are established and product ideas emerge. There has been the tendency in the product-orientated networks to build innovation on technical capacity as a result of this. In large part this may be attributed to the short-term priorities in the Networks for employment creation. The link between employment creation and innovation has served to constrain the projects developed in the Networks and led to a concern with commercial viability, and the attendant problems of marketing and manufacture. The economic environment limits the potential for the production of alternative technologies in alternative ways; given these constraints the Networks have sought to identify the factors that are important prerequisites for the encouragement of user involvement in product development. It was found that information about alternatives was a necessary area to address, together with access to technical facilities for training; for example, the LNTN ran a course for women on electronics. The ability to generate alternative technical solutions to user problems is aided by users understanding the technical hardware; in the Utopia Project, for example, workers were more able to generate alternative ideas about typesetting design specifications once they were familiar with the existing design of the technology. A further area identified in the Networks, particularly Thames Technet, as necessary to address was the promotion of wider discussion on technology issues—the relevance of technology to everyday life, what is the nature of an 'appropriate' technology, and the relationship between technology and gender.

Conclusions

Clearly the GLEB innovation initiative was affected by the wider economic and political environment; the constraint on public sector spending meant that the possibility of influencing public sector demand for socially-useful products was minimal. The difficulty of operating within the competitive market mechanism was demonstrated by the problems of product development in the Networks. The Technology Centres were geared towards the market in areas of high-technology development. It is perhaps possible to explain why this area of innovation support may be more effective by reference to Ray's (1985) criteria for new product assessment. New products must correspond with user needs, complement existing technologies, have some advantage over competing technologies and exhibit these characteristics at an appropriate price. In the Technology Centres products were in line with the existing dominant technological trajectory of high technology, and it appears that this is a crucial factor for access to finance for further product development.

But the establishment of an alternative technological paradigm for the development of socially-useful products required massive resources that are unavailable to local authorities. The GLEB experiment demonstrates the power of market forces for innovative activity; this supports the view that the problem of technology is a political problem. The technology networks began to show how technology could be more relevant to community needs. It appears that socially-directed innovation policy is a question of 'feeling the way'.

Notes

1. It is hard to assess how many jobs were created specifically by the innovation programme since this formed part of the GLC and GLEB's wider investment package. According to GLEB:

> IN THE FIRST two and a half years, to the end of September 1985, GLEB received a total of £60 million in funding from the GLC, including £8 million in mortgage funds.
>
> Of this total, £18 million was invested in enterprises, creating or saving 2,600 jobs, due to rise to nearly 3,000 over two years. A further £16 million was invested in GLEB's property portfolio which, when development work is completed, will provide space for 4,000 jobs, and generate the equivalent of 3,500 years of work in the construction industry.
>
> Over £4 million was allocated to technology networks, and nearly £2 million went on grants and a purchase of fixed assets. Overhead costs totalled £10 million – a lower proportion of budget than most other public sector development agencies. Cash balances totalled £8.5 million, of which £5 million was held against future commitments. ['Enterprising London', GLEB April 1986.]

For further discussion of the jobs issue see the Appendices. A schematic representation of the programme is given in Figure 3 (p. 77).

6 Implications

The Limits of Innovation

The basic question which this book seeks to ask is whether public intervention in the technological innovation process can be used as a tool of social policy. We have looked critically at the conventional, commercially-orientated, approach to innovation and suggested that the market mechanism is often a poor guide as to how to direct innovation to meet social needs. While market-led technological innovation clearly does represent a major mechanism for meeting the goals of private corporations, in terms of increased returns on capital invested from increased productivity and possibly increased market shares based on new products, benefits are not always passed on to either workers or consumers, and their needs are not always reflected in corporate decisions made about technological developments.

Of course few people would wish to dispense with the benefits that technological innovation can, in principle, bestow on society. New products and processes can help us use material resources and skills more efficiently and can help meet specific human needs more effectively.

But these outcomes are not automatic—it is a matter essentially of political struggle and political choice. As Brooks (1973, p. 251) has put it: 'the choices themselves are political, depending upon a complex interplay of bargaining process among conflicting economic, political and ideological interests and values'.

'Radical' local authorities in the UK like the GLC have attempted to intervene in this situation—in part because of the present Conservative government's evident lack of interest in economic intervention. In practical terms, as we have seen, the GLEB, for example, has managed to identify and support the development of a number of new products which they felt were socially relevant. But there have been problems in achieving market deployment given GLEB's limited resources and the dominance of competitive market forces.

Developing a few new products, while perhaps a useful activity in itself (in terms of meeting specific needs), does not represent a substantial contribution to challenging the market's hegemony. That said, the actual development of practical new socially-relevant technologies does act as a sort of technological 'agit prop'—highlighting the technological possibilities and social and political choices.

The GLEB and the technology networks have thus made somewhat more impact as centres for the promotion of practical ideas about alternative lines of technological development to those that have emerged under the influence of market forces. For example, they have pioneered work on human-centred production, and have sought to intervene in other areas of national-level technology policy, in particular, energy policy in relation to combined heat and power and the efficient use of energy.

So while, given the lack of financial resources, it may have proved to be hard for GLEB to challenge the market in material economic terms, it has been possible to put some alternative technological trajectories on the political and intellectual agenda. To go further than that would require significantly larger resources of financial and political power than are available to local councils. As it is, GLEB has found that it has often been constrained by the market to follow conventional lines of technological development in its practical investments—with the more divergent work being threatened with marginalisation as long-term research.

The Future

Some people see bodies like GLEB as essentially prototypes of socialist industrial planning departments in a new Labour government in the UK, with the implication that then they would have sufficient resources to mount a serious challenge to market forces. GLEB itself sees things somewhat differently. While there would be much to be gained by having GLEB-type organisations established locally in each area of the country, they feel that a very different type of institution would be required at national level—even if the policies adopted were similar. But either way the political aim would remain the same: to promote production for need as against production for profit. The question that then emerges is what the extent of this challenge to the market should be.

Are we still talking of a mixed economy in which the private and public sectors, while linked, develop separately, with either peaceful coexistence or permanent running battles as the norm? Or are we talking about the gradual (or rapid) demise of the private sector with some form of social ownership and control taking over? This, of course, has been a continuing debate in British politics often couched in terms of the need for control of the 'commanding heights' of the economy, via nationalisation, as opposed to nationalisation simply of those industries and services deemed to be unprofitable by the private sector.

We have seen that the sort of social control implicit in the local council initiatives has been very much decentralised, local-level control—with investment being channelled to meet local needs, albeit within an overall strategic framework. What has emerged has been the beginnings of 'two-way' top-down and bottom-up planning and participation system.

So conventional centralised 'nationalisation' may not be seen as the way ahead—certainly not across the board, if for no other reason than that the popular experience of the operation of nationalised services has not always been a positive one: they can be bureaucratic and insensitive.

The idea of the full socialisation of the British economy is, of course, only promoted by some people on the Left. A more moderate—one might say reformist—line, seems to have been adopted by the GLC. That is hardly surprising in that the GLC and other councils were in no position to impose full socialisation. Instead some GLC theorists have proposed a new 'hybrid', presumably transitional, approach. As the Chief Economic Adviser to the GLC, Robin Murray (1985), has put it: 'In intervening in the market economy, what we aim to do is to strengthen the socially useful forces, however bound they may finally be by the conditions of profitability. That is why we speak of operating in and against the market.'

The strategy is thus one of challenging the market where possible from within by giving sustenance to the countervailing forces for socially-useful production. While this may be appropriate (and indeed the only feasible option in practical terms) for the radical local authorities, it is far from clear whether it would be sufficient at the national level, where one might expect an even more challenging strategy might be feasible in relation to social and industrial planning and private financial interests.

At the very least constraints would need to be put on the way the market mechanism operates—and not just by intervening more effectively with more resources, via progressive investments in socially needed technology.

Murray (1985) sees it in terms of economic and industrial policy: with Keynesanism being replaced by progressive planned sectorially targeted industrial investment: 'We see industrial policy as being primary, conducted not through the general manipulation of markets, but by particular interventions in production, enterprise by enterprise, branch by branch'. But he adds that 'as long as the market and profitability remain the main economic mechanisms there will be an inherent tension', with the implication that, ultimately at least, these mechanisms should be at the very least augmented by other, more socially-determined decision-making mechanisms.

It is not the purpose of this book to explore all the implications and problems facing the sort of socialist policies implied by such suggestions, for that would take us well beyond our focus on innovation policy. However, we can at least perhaps see from the GLC/GLEB experience the extent to which the strategy adopted by the GLC/GLEB came against the limits of the convential British welfare socialism approach. Whether a state-level initiative would transcend these limits remains to be seen.

After all, it should be pointed out that the GLEB approach was very much tailored to the local and regional level. While it is hoped that its activities illustrated the role that intervention by public agencies like local authorities could play in job creation and economic restructuring, it recognised that different institutional structures would be required at national level. Economic restructuring could not be achieved nationally just be setting up GLEB-type organisations locally; there would be a need for national-level co-ordination and planning by central government.

Some commentators have suggested that the GLC/GLEB initiative has not just been hampered by lack of such national-level planning but has been actually undermined by the government's policies.

The Labour group's original manifesto had commented that 'although national measures will also be necessary, there is much that a regional authority like the GLC could do' (Barratt-Brown 1984, Part I). However, Michael Barratt-Brown (1984, Part III) has argued that in the event the GLC's overall programme of investing in jobs and restructuring London's Economy 'was in effect rendered impossible by the absence of any positive national measures, indeed by the intensification of Government cuts in services, by privatisation of the public sector and by the capping of Local Government rates and ultimately the abolition of the GLC'. He suggests that much of the GLC's industry and employment work was inevitably therefore defensive—trying to resist cuts.

Now this may be true of some aspects of the GLC industry and employment work and of the work of the Popular Planning Unit in particular. After all, the GLC believed that what Londoners most needed in many cases was the re-establishment of the services (health, transport and housing) that were being cut. But the GLEB did seem to have been able to initiate successful economic interventions via its sector planning, employment and innovation programmes, from which several lessons can be learnt.

The Lessons

Perhaps the most obvious first point to make is that innovation is a slow and often erratic process. Even given commercial incentives, it takes time and is not guaranteed to deliver. Attempting to direct innovation to social ends is likely to be even harder. Given these problems, there is always the temptation to adopt a 'technology-led' approach—to assume that all that is needed is to provide funding for technical researchers to push ahead with technically exciting and developing projects. As we have seen, the GLEB technology networks have learnt to ameliorate this tendency by involving themselves with local community issues and organisations, so as to introduce a strong 'need-led' element into the process.

In a sense they have moved more to the 'Science Shop' model of community problem-directed research operating in Denmark, although they also have the benefit of broader, more strategic sectoral overviews provided by the GLEB.

What has emerged, then, is a two-way process of 'local need identification' by the Networks (and the GLC Popular Planning Unit until its demise), coupled with strategic sectoral analysis and planning by the GLEB (and, until 1986, the GLC's Economic Policy Group) with the GLEB applying social and strategic criteria to any subsequent investment.

We have seen that this process does not always work smoothly: the various agencies involved may be working on different time-scales and with differences in operational styles. And we have seen that the next stage—the move to production—has proved to be difficult.

That introduces the second key point. 'Need-led' innovation has to be linked with the necessary production resources. The GLEB has increasingly found it necessary to try to link product (and service) ideas from the networks to GLEB-supported or linked companies.

The implications are that there is a need for a degree of municipalisation of production—given private capital's evident lack of interest in socially-needed products. The GLEB might thus be seen to be reflecting, at the local level, the ideal of socialisation of production which Labour governments have considered (and carried out in some sectors) at national level.

However the analogy cannot be carried too far. The GLEB does not have the resources to fund full-scale production—all it can do is try to provide pump-priming investment to support selected private companies to launch socially-relevant products.

There is clearly a risk in this approach that, unless the public agency negotiates strict controls, private companies may seek to ignore some or all of the social criteria (for example, concerning methods of production) and simply exploit the technical advantages in their own corporate interest. Strong trade union involvement is obviously required in order to police the situation.

Similar problems exist in relation to protection of the product. The publicly funded product innovations developed via the GLEB could be exploited simply for commercial gain unless the GLEB imposed some protective licensing arrangements to ensure that benefits would return to the community. The GLEB would have to make an assessment of job gains and other benefits accruing from commercial exploitation before letting go of any of its developments. The GLEB would thus be using the innovations it had developed as bargaining tools in an attempt to shape the pattern of subsequent manufacture and marketing policies adopted by the private companies it dealt with.

Whether that is a viable control option will depend to a great extent on the attractiveness of the product, that is, how keen companies are to get hold of it.

The experience so far has not always been encouraging—as is shown by the examples of Pedelec and Airlec. Then there is the risk that once the GLEB has been forced to abandon such projects, due to lack of interest from the private sector, private capital will pick them up, thus avoiding any GLEB influence on subsequent developments (this seems to have happened with Airlec and Pedelec). Clearly the GLEB could guard against this by taking patents or some other form of protection out for the product idea—but that might frighten off potential investors interested only essentially in what amounts to asset stripping.

There could be a similar problem in relation to the companies set up initially with GLEB support. As they mature and pay back any debts to GLEB, they may move away from their parent and adopt different policies. Certainly GLEB's direct influence could diminish, even if the companies remain faithful to the basic principles, unless the GLEB maintains some sort of representation on the board. (In this context it will be interesting to see what happens to Whitechapel Computers which has proved to be a very successful company developing some novel and well-received computer systems).

The general lesson is that it is evidently likely to be hard to operate 'in and against' the market by using investment in new products and companies to influence subsequent developments; 'what seems to be important is, first, the licensing and control arrangements that are negotiated between the GLEB and the enterprise, and second, the involvement of the workforce to monitor subsequent company policy and development in line with the negotiated Enterprise Plan'.

The third point concerns the general direction of innovation. Expediency (for example, in relation to job creation) may dictate that investment be channelled into technologies which can be developed rapidly and deployed easily. In particular, it may be tempting to opt for the areas of development currently favoured by financiers and investors: bowing to profit-led criteria may appear inescapable if full-scale production is the aim. Even if public funds are available; it is hard to take technological risks with public money. Thus while it might be hoped that 'need-led' innovation would produce more socially appropriate and therefore more genuinely successful innovations, in the struggle to succeed, there is always a temptation to adopt a technological 'tunnel-vision' approach and thereby ignore possibly important areas of development. And yet little will have changed unless this temptation is resisted.

One of the key potential advantages of the socially-directed innovation approach is that hopefully a more diverse pattern of innovation can be stimulated with new directions being opened up.

Given the dominance of conventional approaches to innovation, and the widespread acceptance of existing technological trajectories, it may seem too much to expect beleaguered local councils, community and trade union organisations to pioneer alternative approaches to technology. But as things stand they are amongst the few agencies in society that have an interest in doing that. Whether this will remain the case will to some extent depend on how widely their ideas and experiences are diffused and accepted.

So perhaps the crucial point concerns popular communication. In the end it is the public who, via taxes and rates, invest in the sorts of project we have discussed. It is vital that they be kept fully informed of how their money is being spent as they are essentially shareholders with democratic rights to direct policy.

The whole point of the exercise is for their needs to be paramount. That, of course, implies that local consumers, users of services, and so on, also have a responsibility—to indicate their needs and to involve themselves in local organisations which can transmit them effectively.

Socially-directed innovation must be just that—socially directed—or else it will become just another technological fix.

Conclusions

The analysis presented can be summarised as follows. Firstly, there exists a conceptual polarity between technology-push (with all its connotations of technological determinism) and market-pull (with its links to commercial values rather than social needs). Although it reality these two influences interact in the innovation process, many people look to technology-led innovation as the solution to our economic (and by inference, social) problems.

Some radicals, when approaching innovation, have, initially at least, adopted a similar stance, simply replacing profit-led innovation with the idea of 'socially-useful' products as pioneered by the Lucas Aerospace workers. The danger with this approach is that it may reduce to simple technological determinism—products become reified and 'abstracted' from their context despite the rhetoric of 'social usefulness'.

This danger has usually been avoided, however, in that, in the case of the GLEB, for example, the emphasis was always on need-led innovation, with the social utility criteria spelt out: the unification of 'product' and 'context' discussed in Part One was at least attempted.

Even so the initial emphasis was often on 'new products' in the belief that it was strategically important—and possible—to intervene in the early stage of the innovation process, that is, the invention/new product idea stage. That point of intervention was felt to offer the opportunity for some influence over the way the innovation process developed subsequently.

The reality has proved to be somewhat different: as we have seen, moving from invention to manufactured and distributed products takes time and is fraught with many difficulties. It is hardly surprising, then, that many of the radical councils backed off from this approach. Roughly speaking, the GLEB and the technology networks were the most ardent adherents, followed by Sheffield (and SCEPTRE) and then WMEB (and UDAP). WMEB and later Sheffield City Council, maybe to a different degree, viewed the 'socially-useful product' idea as a long shot—and tended to put more resources into diffusion of information about existing products, and this tendency has strengthened, with a growing emphasis on 'technology transfer' from existing institutions via conventional commercial Science Parks.

The GLEB by contrast maintained its commitment to socially-useful technology via the technology networks, even if, under the pressure of its reduced circumstances after the abolition of the GLC, it has not been able to maintain a high level of support for this approach.

One question that emerges is whether this shift of emphasis represents a retreat. Certainly some radicals view support for commercial Science Parks with concern. But it might also be argued that, while that particular development may be problematic, a shift in emphasis away from 'new products' towards an assessment of needs is actually a radical move. For example, as we have seen, LEEN, the energy network, argued from its experience that the problems they identified often did not require 'new products' for their solution, but rather access to existing technologies. To put it simply, the emphasis moved from radical socially-useful technology-push to radical need-pull.

Now, of course, there are just as many problems with operating at this end of the innovation process: 'needy' individuals and groups may not be able to articulate their 'needs' as economic demands since they do not have the financial resources. And while focusing on needs may throw up ideas about how technology should be developed and deployed, the technologies that are available are those that have been developed by and for the conventional commercial systems.

It would seem that some sort of balance between technology supply (development) and technology use (social context) is still needed. Whether that would avoid, on the one hand, the difficulty experienced in attempting to exert social control over the process of technological development and, on the other hand, contribute to the process of socialising the market mechanism, remains to be seen.

Perhaps all that can be concluded at the present stage is, first, that some innovative and socially-needed products can be developed in some sectors; second, that in some sectors it is possible to identify and, in a limited way, support 'social markets' for some new or existing products or services; and third, that these activities present at least an element of a challenge to conventional approaches and beliefs about both innovation and the role of the market in matching technology to needs.

7 General Conclusions

The experiments we have looked at into the development of alternative technological trajectories highlight the difficulties of operating outside the dominant technological paradigm. As we have seen, the concept of technological paradigm (Dosi, 1982) can be used to explain how the dominant routes of technological development emerge, and the fact that the selection process is shaped by powerful economic, political and institutional factors illustrates the political nature of innovative activity and challenges the idea of a technological determinism. The choice of specific types of 'high-technology' as the appropriate area of current development in Britain, and its attendant association, in ideological terms, with technological progressiveness, acts to exclude and limit technological development in areas that are considered 'low-technology' or in some other way, as inappropriate. This may be seen as a contributory factor to the GLEB adopting a 'high-technology' approach in the Technology Centres and increasingly marginalising the support for the development of socially-useful products in the Technology Networks. Does this mean then that the insights into the social construction of technology are of little use practically?

Noble (1983a) argues that the missing ingredient in the construction of alternative technologies is power. Workers and users do not have access to the political and economic resources with which to secure the production, manufacture and diffusion of alternative technologies. This was evident with the alternative product proposals that emerged from the Lucas Plan, and to a large extent was similarly the case with the GLEB experiment. As an exercise in innovation in a competitive market economy, economic and commercial viability are the assessment indicators—and in these terms the experiment was largely a failure. But does that matter? The political point after all was to criticise the existing technological trajectories, and demonstrate the possibilities of alternatives.

The GLEB experiment highlighted the assessment criteria for the selection of routes to alternative forms of technological development, and attempted to substitute the profit motive with 'social use' criteria. This social process questions the notion of economic viability on the market. The notion of economic viability is a political construction, so that alternative technologies might be economically viable if there were the political commitment of resources to their development. Once it is established that change in the political priorities of governments (and companies) is crucial, then the problem of how to effect that change emerges. The GLEB experiment demonstrates the importance of political mobilisation around constructive and practical alternatives, i.e. practical demonstrations of alternatives, which illustrate an alternative paradigm.

Essentially the alternative paradigm is one in which social 'use value' replaces economic exchange value, to use Marxist terminology, with the emphasis on the social factors surrounding the *linking* of production and use. In terms of production the focus is thus for example on human centred production systems rather than on de-skilling. In terms of consumer products and services, the emphasis is on ensuring that a wide range of social groups can influence the innovation process. A number of specific criteria emerge from this process—concerning social and environmental priorities and concerns—and they can be refined and developed, enriching the paradigm and aiding selection of appropriate technological trajectories. (Elliott 1986).[1]

The alternative paradigm can be developed piecemeal to some extent—as a subculture within the dominant paradigm, although clearly there are limits to how far this can proceed. At some point conflict is inevitable: that now seems to be the case in relation to nuclear versus alternative energy. But in the end the conflict is not over specific technologies: rather it concerns the analytic frame of reference, social priorities and criteria. Ultimately it is a paradigm conflict.

As we have suggested, alternative paradigms of this type serve to delineate a field of enquiry that is based on the social context of technology: the solutions to problems and the questions that arise originate from the producers and users at the implementation end of the innovation process. The concentration on the context of use, environmental effects and cultural effects of technologies gives rise to questions that can lead to a more appropriate route of technological development. For example, does the technology have advantages in terms of use, what are the consequences for individuals of its introduction, are there damaging ecological effects, will the introduction of the technology affect the organisation of work, gender relations and the wider social system? This ought to be a *practical* activity: the development and assessment of alternative product ideas requires a process of design-by-doing, as was exemplified by the

UTOPIA project. Furthermore, this process can also involve adoption and modification of some existing products rather than the development of totally new ideas.

The concentration upon practical technological activity is a key area for the development of alternative technologies. The innovation spiral discussed in Part One indicated the cyclical nature of the innovation process—products are—or should be—invented in relation to a need, produced and diffused and then re-adapted and re-designed along a particular trajectory that is influenced by market demands. The stage of re-adaption and re-design may in many cases be taken as the starting point since the recognition of a need for an alternative is context dependent and related to the unworkability of technologies in the social setting. For example, the use of public transport generates the idea for the need for changes in the mounting platform—for women with young children and heavy shopping and for the elderly or infirm the step up is too high. For the generation of creative popular design, the key point of intervention in the innovation process in this area of re-design. The workers in project UTOPIA found that they were able to generate alternative solutions for the design specifications of type-setting equipment, once they were familiar with the hardware.

Training programmes in technological awareness and the creation of facilities for working with technology and generating alternative solutions is an area that can be addressed by local authorities. The Technology Networks are a good example. As we have seen, there is a clear need to support both practical training and the more general process of creating space (and time) for the discussions of technology issues, if alternative needs were to be articulated by workplace and community groups. Whether practical awareness or conceptual/social understanding is the key factor in defining *need* is of course debateable: there is an interaction between 'vision' and day to day tacit knowledge and experience. But time and resources are needed in order for this interaction to develop.

Socially-directed innovation requires a hegemonic strategy that influences attitudes to technology, and promotes a popular technological awareness. Practical exemplary projects have a role as a focus for political mobilisation. The coupling of technology with effective demand is crucial, in this case effective demand is political power.

The political framework of a local authority can act as a base for political mobilisation around technology issues, what is necessary is a coordinated strategy. If the GLEB experiment is seen in terms of Callon's (1980) notion of the 'actor world' necessary to the development of technologies, then it is possible to see their initiative as a stage-set or scenario. The development of a scenario for socially-useful product development required the assignment of roles incorporating all the aspects necessary for innovation within an overall framework. There were a number of difficulties with the GLEB scenario or actor world that led to its breakdown. Crucial components were missing—requisite social groups to articulate needs, attention to demand factors, lack of co-ordination and perceptional differences between different groups involved in the political/technological project. These points have been discussed earlier, what they indicate is the difficulty, with policy making, of what comes first. The GLEB experience shows the need for a clear statement that can act as the focus and link between policy and implementation.

Alternative technological development is a learning process that is influenced by internal and external factors. The possibility of affecting external factors, that is, the wider political and economic environment, is enhanced by internal co-ordination.

Everyday life is dependent on technology: but the increasing specialisation of the innovation process and technological development serves to remove its relevance to lay people. While the possibility of influence on the technological environment can appear remote, the alternative experiments as we have discussed and the theoretical insights into the social construction of technology that we have tried to develop, highlight the close practical and political relationship that could be created between technology and society. Opportunities do exist to exert influence: the practical experiments we have looked at give an indication of the points that need to be addressed—the promotion of technological awareness, alternative social marketing techniques, education and training facilities. Their value and role within a hegemonic counter-strategy to existing routes of development is crucial.

Notes

1. The alternative paradigm is spelt out in more detail in Elliott D. (1985). 'Technology and the future' technology policy group occasional paper no. 8, Milton Keynes, The Open University. A version of which also appears in the *Journal of Interdisciplinary Economics* (1986 Vol. 1).

Postscript

As we go to press, GLEB is still struggling to continue its work—albeit on a somewhat reduced basis, with the support of twelve of London's thirty-two local councils. However, with significantly less money available, staff redundancies have been threatened. So far, forced redundancies have been avoided and, although some staff have left, GLEB continues to operate without a major shift in policy. How long this can continue, of course, remains to be seen.

Many activists believe that the future of GLEB and the other council projects could only really be ensured if a sympathetic government is returned at the next general election, in 1987 or early 1988. Certainly, as has been noted, the Labour Party has put support for GLEB-type schemes high on its agenda, while the Conservative Party on 26 November 1986 launched its own plan for a network of commercially-orientated regional Technology Centres.

Appendices

1. Documentation

In its last year of operation the GLC produced two major reports which amongst other things, serve as summaries of the various projects we have reviewed. These were 'The London Industrial Strategy' (GLC 1985) and 'The London Technology Strategy' (GLC 1986) The latter has since (1987) been published independently by Comedia (9 Poland Street, London W1) under the title *Britains Industrial Renaissance?* (eds P. Blackburn & R. Sharpe). They should also be available from GLEB, 63-7 Newington Causeway, London, SE1, as are a wide range of publications on GLEB's work (01-403-0300).

In addition eight London boroughs have supported the establishment of the London Strategic Policy Unit (LSPU) which will continue some of the GLC sector analysis and research activities. LSPU can be contacted at Middlesex House, 20 Vauxhall Bridge Road, London SW1V 2SB (01-633-3710).

2. The Five TechNets

London Energy and Employment Network (LEEN): 99 Midland Road, NW1 ZAH, (01-380-1002).

London Innovation Centre: Unit B, Hornsey Street, N7 8HR, (01-607-8141).

London New Technology Network (LNTN): 86-100 St. Pancras Way, NW1 9E5F, (01-482-3816).

London Transport Technology Network (Transnet): c/o GLEB or Thames Technet.

Thames Technet: 16 Warren Lane, Woolwich, SE18, (01-854-2511).

3. Employment

By the beginning of 1984 GLEB had invested £20 million in some 173 enterprises and projects, directly creating or saving 2,000 jobs and providing space via its property investments for a further 4,000 jobs plus 2,500 construction jobs. In addition it had spent £3 million setting up the Technology Networks which were hoped ultimately to lead to even greater job gains.

But the number of jobs and the speed with which they could evidently be created given GLEB's approach is not the only point. Equally significant was the relative *cheapness* of the job creation exercise – at some £10,000 per job place directly saved or created. As GLEB commented in a Press Release (22 November, 1983):

> The cost of each job created or saved by the GLEB is about 15 per cent of the cost of jobs generated in Government supported enterprise zones. According to one calculation, published in the *Financial Times* and based on government figures for spending on infrastructure projects in the zones, and the cost of tax and rates concessions, each job generated within the enterprise zones to the end of last year cost more than £37,000. But even this may be an under-estimate since many of the jobs relate to firms which have merely re-located inside the enterprise zones. If these jobs are excluded the average figure rises to no less than £94,000.

This calculation comes out even more favourably for GLEB if the focus is narrowed to just the direct 'enterprise' investments of £8 million in 1983/84 for the 2,000 jobs saved or created – £4,000 per job.

Subsequently with the total direct jobs created or saved approaching 3,000 even more favourable figures were claimed. In March 1985, GLEB's independent auditors, Thornton Baker, calculated that the cost per job was falling in the range £2,700–£7,400 per job, depending on the contract and accounting assumptions.

The *New Statesman* (25 March, 1985) commented: 'the most realistic assumptions produce a figure of about £3,500 per job – half the average cost to the taxpayer of an unemployed person in benefits and lost taxes'.

Bibliography

Alvey, J. (chmn) (1982), *A Programme for Advanced Information Technology*, The Report of the Alvey Committee, London, HMSO.

Athanasiou, T. (1985), 'High-Tech Alternativism: The Case of the Community Memory Project', *Radical Science Journal*, no. 16, January.

Barratt-Brown, M. (1984), *The Greater London Enterprise Board*, Social and Economic Study Pack no. 3, London, GLC Economic Policy Group.

Barratt-Brown, M. (1986), *London and the Third World*, Social and Economic Study Pack no. 6, London, GLC Economic Policy Group.

Bell, D. (1974), *The Coming of Post-Industrial Society: A Venture in Social Forecasting*, London, Heinemann.

Bennington, J. (1986), 'Local Economic Strategies', *Local Economy*, no. 1, spring.

Boddy, M., (1984), 'Local Economic and Employment Strategies' in M. Boddy and C. Fudge (eds), *Local Socialism?*, London, Macmillan.

Bodington, S., George, M. and Michaelson, J. (1986), *Developing the Socially Useful Economy*, London, Macmillan.

Braun, E., (1984), *Wayward Technology*, London, Frances Pinter.

Braverman, H., (1974), *Labor and Monopoly Capital*, New York, Monthly Review Press.

Brooks, H. (1973), 'Technology Assessment as a Process', *International Social Science Journal*, no. 3.

Brooks, H. *et al.* (1971), *Science, Growth and Society: A New Perspective*, Paris, OECD.

Brooks, H. *et. al.* (1974), 'Technology Assessment and Technology Forecasting', in N. Cross, D. Elliott and R. Roy (eds), *Man-Made Futures*, London, Hutchinson Educational.

Bruce, M. (1984), 'Technology Assessment of Interactive Videotex', PhD thesis, Manchester Polytechnic.

Bruce, M. (1985), 'A Missing Link: Women and Industrial Design', *Design Studies*, vol. 6, no. 3, July.

Bruce, M., Kirkup, G. and Thomas, C. (1984), *Teaching Technology Assessment to Women*, Milton Keynes, Open University Design Discipline.

Burawoy, M. (1978), 'Toward a Marxist Theory of the Labour Process: Braverman and Beyond', *Politics and Society*, vol. 8, no. 2.

Callon, M. (1980), 'The State and Technical Innovation: A Case-study of the Electric Vehicle in France', *Research Policy*, no. 9.

Carter, E. (1985), 'Alternative Products in West Berlin' in Collective Design/Projects (eds), *Very Nice Work If You Can Get It*, Nottingham, Spokesman.

City of Sheffield Employment Department (1982). *Report*; Sheffield City Council.

Cockburn, C. (1981), 'The Material of Male Power', *Feminist Review*, no. 9.

Collins, H.M. (1983), 'The Sociology of Scientific Knowledge: Studies of Contemporary Science', *Annual Review of Sociology*, vol. 9.

Cooley, M. (1983), *The New Technology—Social Impacts and Human Centred Alternatives*, Milton Keynes, The Open University Technology Policy Group, Occasional Paper no. 4.

Cooley, M. (1984), 'Socially-useful Design' in R. Langdon and N. Cross (eds), *Design Policy, Vol I: Design and Society*, London, Design Council.

Cooley, M. (1985), 'After The Lucas Plan', in Collective Design/Projects (eds), *Very Nice Work If You Can Get It*, Nottingham, Spokesman.

Dosi, G. (1982), 'Technological Paradigms and Technological Trajectories', *Research Policy*, no. 11.

Doyal, L., and Gough, I. (1984), 'A Theory of Human Needs', *Critical Social Policy*, no. 10.

Editorial Collective (1982), 'A Socialist GLC in Capitalist Britain?', *Capital and Class*, no. 18.

Ehn, P., and Kyng, M., (1985), 'The Unions and Computers: The Scandinavian Collective Resource Research Strategy', Brussels, EEC FAST Conference Paper.

Elbaum, B. *et al.* (1979), 'The Labour Process, Market Structure, and Marxist Theory', *Cambridge Journal of Economics*, vol. 3, no. 3.

Elliott, D. (1986), 'Design Criteria and Innovation: Some Conclusions', *Design and Innovation* (T362), Unit 16, Milton Keynes, Open University Press.

Employment Programme Committee (1983), *Report*, 'Strategies for the Employment Department 1983/4', Sheffield City Council.

Freeman, C. (1979), 'The Determinants of Innovation', *Futures*, vol. 11, no. 3.

Freeman, C. (1982), *The Economics of Industrial Innovation*, 2nd. edn, London, Frances Pinter.

Freeman, C., Clark, J. and Soete, L. (1982), *Unemployment and Technical Innovation*, London, Frances Pinter.

Galbraith, J.K. (1969), *The New Industrial State*, Harmondsworth, Penguin.

Gershuny, J. (1985), 'New Technology—what new jobs?', *Industrial Relations Journal*, Autumn.

Gibbons, M. and Johnston, R.D. (1970), 'The Relationship between Science and Technology', *Nature*, no. 227.

Goodwin, M. and Duncan, S. (1986), 'The Local State and Local Economic Policy', *Capital and Class*, no. 27, Winter.

GLC (1982), Industry and Employment Committee Report 413, London, GLC.

GLC (1983), *Jobs for a Change*, London, GLC Economic Policy Group.

GLC (1984a), *Jobs for a Change*, no. 11, London, GLC Popular Planning Unit.

GLC (1984b), *Jobs for a Change*, no. 13, London, GLC Popular Planning Unit.

GLC (1985), *The London Industrial Strategy*, London, GLC.

GLEB (1983), *Enterprise Planning*, London, GLEB.

GLEB (1984b), *Enterprising London*, no. 2, September, London, GLEB.

GLEB (1984d), Press Release no. 47, London, GLEB.

GLEB (1984a), *Enterprising London*, no. 1 May, London, GLEB.

GLEB (1984c), Press Release, no. 44, London, GLEB.

GLEB (1985a), *Enterprising London*, no. 4, March, London, GLEB.

GLEB (1985b), *Enterprising London*, no. 5, Summer, London, GLEB.

GLEB (1985/6), *Corporate Plan*, London, GLEB.

Haeffner, E.A. (1973), 'The Innovation Process', *Technology Review*, March/April.

Howard, R. (1985), 'UTOPIA: Where Workers Craft New Technology', *Technology Review*, April.

Hughes, T.P. (1986), 'The Seamless Web: Technology, Science Etc. Etc.', *Social Studies of Science*, vol. 16, no. 2.

Isensen, R.S. (1967), 'Project Hindsight' in Kelly and Kranzberg (1978).

Johnston, R. (1985), 'The Social Character of Technology', *Social Studies of Science*, vol. 15.

Kelly, P. and Kranzberg, M. (eds) (1978), *Technological Innovation: A Critical Review of Current Knowledge*, San Francisco, San Francisco Press.

Kling, R. (1980) 'Social Analysis of Computing: Theoretical Perspectives in Recent Empirical Research', *Computing Surveys* 12 (1) March 1980.

Knorr-Cetina, K. and Mulkay, M. (eds) (1982), *Science Observed*, Beverly Hills/London, Sage.

Kuhn, T. (1962), *The Structure of Scientific Revolutions*, Chicago, Chicago University Press.

Langrish, J. *et al.* (1972), *Wealth from Knowledge*, London, Macmillan.

Lawless, E. (1977), *Technological Shock*, New Brunswick, NJ, Rutgers University Press.

Layton, E. (1974), 'Technology as Knowledge', *Technology and Culture*, vol. 15.

Lazonick, W. (1979), 'The Self-acting Mule and Social Relations in the Workplace', in MacKenzie and Wajcman (1985).

Liff, S. (1985), 'Analysis of West Midland Approaches to Technology Transfer', Aston University, Technology Policy Unit.

Lowe, J. (1985), 'Science Parks in the UK', *Lloyds Bank Review*, April.

MacKenzie, D. and Wajcman, J. (eds) (1985), *The Social Shaping of Technology*, Milton Keynes, Open University Press.

Mansfield, E. (1968), *Industrial Research and Technological Innovation*, New York, Norton.

Marvin, S.J. (1986), *High-hopes for High-Tech: Local Authority Technology Policies and Initiatives: An Overview*, Milton Keynes, Open University Technology Policy Group Occasional Paper no. 13.

Minns, R. (1982), *Take Over the City*, London, Pluto Press.

Moreton, A. (1983), 'Science Parks', *Financial Times*, 21 January 1983.

Mowery, D. and Rosenberg, N. (1979), The Influence of Market Demand upon Innovation, *Research Policy*, no. 8.

Mulkay, M.J. (1979), Knowledge and Utility, *Social Studies of Science*, vol. 9.

Murray, R. (1985), 'London and the GLC: restructuring the Capital of capital', *IDS Bulletin*, University of Sussex, vol. 16, no. 1, January 1987.

Myers, S. and Marquis, D. (1969) *Successful Industrial Innovations*, Washington, DC, National Science Foundation, Report NSF 69-17.

NEDO (1983), *Civil Exploitation of Defence Technology*, London, National Economic Development Office, Electronics EDC.

NEC (1986), *Statements to the 85th Annual Conference of the Labour Party*, Blackpool, Labour Party.

Nelson, R. and Winter, S. (1977), 'In Search of a Useful Theory of Innovation', *Research Policy*, no. 6.

Noble, D. (1979), 'Social Choice in Machine Design: The Case of Automatically Controlled Machine Tools' in A. Zimbalist (1979), *Case Studies in the Labor Process*, New York, Monthly Review Press.
Noble, D. (1983a), 'Present Tense Technology', *Democracy*, vol. 3, no. 3.
Noble, D. (1983b), 'Present Tense Technology', *Democracy*, vol. 3, no. 4.
Palmer, J. (1986), 'Enterprise Boards: A Growing International Movement' in 'Back to the Future', Supplement to *New Socialist*, March.
Papanek, V. (1974), *Design for the Real World*, St Albans, Paladin.
Pelly, D. (1985), 'Arms Conversion and the Labour Movement' in Collective Design/Projects, *Very Nice Work If You Can Get It*, Nottingham, Spokesman.
Pinch, T.J. and Bijker, W.E. (1984), 'The Social Construction of Facts and Artefacts: or How the Sociology of Science and the Sociology of Technology Might Benefit Each Other', *Social Studies of Science*, vol. 14, no. 3.
Pinch, T.J. and Bijker, W.E. (1986), 'Science, Relativism and the New Sociology of Technology: Reply to Russell', *Social Studies of Science*, vol. 16, no. 2.
Polanyi, M. (1976), *The Tacit Dimension*, London, Routledge.
Ray, T. (1985), 'Post-Innovation Performance: The Process of Technological Change at Firm Level', PhD thesis, Manchester Polytechnic.
Rosenberg, N. (1974), 'Science, Invention and Economic Growth', *Economic Journal*, vol. 84.
Rosenberg, N. (1982), *Inside the Black Box: Technology and Economics*, Cambridge, Cambridge University Press.
Roy, R. (1986), 'Introduction: Meanings of Design and Innovation' in R. Roy, and D. Wield (eds), *Product Design and Technological Innovation*, Milton Keynes, Open University Press, 1986.
Russell, S. (1986), 'The Social Construction of Artefacts: A Response to Pinch and Bijker', *Social Studies of Science*, vol. 16, no. 2.
Science (1984), vol. 223, 16 March.
SPRU (1971), 'Report on Project SAPPHO', Brighton, University of Sussex, Brighton, SPRU.
Schwartz-Cowan, R. (1979), 'Gender and Technological Change' in MacKenzie and Wajcman (1985).
Schwartz-Cowan, R. (1983), 'How the Refrigerator Got its Hum' in MacKenzie and Wajcman (1985).
Shapin, S. (1982), 'History of Science and its Sociological Reconstructions', *History of Science*, vol. 20.

Sheffield in the Eighties (1985), 'A Strategy for Industry and Jobs', Sheffield, Sheffield City Council Employment Department.

Sheffield Technology Campus (1985), Presentation Summary, Sheffield City Council.

Stockholm International Peace Research Institute (SIPRI) Yearbook (1986), *World Armaments and Disarmaments*, Oxford, Oxford University Press.

'Struggles in the Welfare State', (1984), *Critical Social Policy*, no. 10, Spring.

Utterback, J.M. (1974), 'Innovation in Industry and the Diffusion of Technology', *Science*, vol. 183.

Von Hippel, E. and Finkelstein, S.N. (1978), 'Product Designs which encourage—or discourage—related innovation by users', cited in Rosenberg (1982).

Wainwright, H. and Elliott, D. (1983), *The Lucas Plan*, London, Allison and Busby.

Walker, D. (1986), 'Design and Innovation: An Introduction', *Design and Innovation* (T362), Units 1–2, Milton Keynes, Open University Press.

Walker, M. (1986), 'Defence Spending' in Roy and Wield (1986). Op. cit.

Ward, M. (1983), 'Labour's Capital Gains: the GLC Experience', *Marxism Today*, December.

Weinberg, A. (1966), 'Can Technology Replace Social Engineering', *The University of Chicago Magazine*, October.

Weizenbaum, J. (1976), *Computer Power and Human Reason*, San Francisco, Freeman.

Wilkinson, B. (1983), *The Shopfloor Politics of New Technology*, London, Heinemann.

Wood, S. (ed.), (1982), *The Degradation of Work? Skill, Deskilling and the Labour Process*, London, Hutchinson.

Wynne, B. (1983), 'Redefining the Issues of Risk and Public Acceptance: the Social Viability of Technology', *Futures*, February.

Index